giving
nature
a home

Ospreys

Tim Mackrill

BLOOMSBURY WILDLIFE

LONDON · OXFORD · NEW YORK · NEW DELHI · SYDNEY

BLOOMSBURY WILDLIFE
Bloomsbury Publishing Plc
50 Bedford Square, London, WC1B 3DP, UK

BLOOMSBURY, BLOOMSBURY WILDLIFE and the Diana logo are trademarks of
Bloomsbury Publishing Plc

First published in Great Britain 2019

A catalogue record for this book is available from the British Library

Library of Congress Cataloguing-in-Publication data has been applied for

ISBN: PB: 978-1-4729-5603-3; ePub: 978-1-4729-5602-6; ePDF: 978-1-4729-5604-0

2 4 6 8 10 9 7 5 3 1

Design by Rod Teasdale

Printed and bound in India by Replika Press Pvt. Ltd.

MIX
Paper from
responsible sources
FSC
www.fsc.org   FSC® C016779

To find out more about our authors and books visit www.bloomsbury.com
and sign up for our newsletters

giving
nature
rspb a home

Contents

Meet the Osprey

The striking brown-and-white Osprey is one of the world's most widely distributed and best-loved birds of prey, with breeding or wintering populations on every continent except Antarctica. The only bird of prey active during the day that feeds exclusively on fish, it has been revered for centuries for its hunting prowess. Northern populations are migratory, with individual birds undertaking long and perilous journeys alone to reach wintering grounds many thousands of kilometres away.

The hunting Osprey has its piercing yellow eyes fixed on ripples in the water below. Suddenly, it folds its wings and drops, arrow-like, at incredible speed. At the last moment it throws its powerful talons forward and crashes into the water. After a few seconds with its wings splayed on the water's surface, it musters the energy to take off again. A few flaps of its vast wings lift it clear of the water, revealing a fish carried head first. A mid-air shake sends a shower of water droplets downward, and then the supreme hunter powers off with its catch. There are few spectacles in the natural world as impressive or dramatic as a fishing Osprey, and if you are lucky enough to enjoy watching one at close quarters it is a sight that lives long in the memory.

Below: *Pandion haliaetus ridgwayi* – the subspecies of Osprey that occurs in the Caribbean – has a much whiter head than its counterparts elsewhere in the world, giving it a striking appearance.

Opposite: Ospreys catch a range of fresh and saltwater fish from on or just below the surface. Once they have a tight hold of their prey in their highly adapted talons, a series of powerful wing beats helps them get clear of the water.

What is an Osprey?

The Osprey (*Pandion haliaetus*) is one of the best-known and most popular birds of prey in the United Kingdom and around the world. You can enjoy watching fishing Ospreys from North America to Australasia, and from Europe to Asia. In fact, this supreme hunter can be found on every continent except Antarctica at some point in the year.

Most northern populations of Ospreys are migratory, with individual birds making long and perilous journeys across oceans and deserts twice a year. Adult Ospreys, which may live into their twenties, often remain faithful to the same breeding and wintering sites throughout their life and, after spending the winter apart, are reunited with the same mate upon returning to their nest each spring. Whether atop a pine tree beside a remote Scottish loch, or a telegraph pole in a car park in North America, a newly returned Osprey sitting resplendent beside its nest is a sure sign that spring has arrived. It is little wonder that this bird is revered just about everywhere it occurs.

Below: Most Ospreys from northern regions migrate south for the winter. Amazingly, juveniles undertake the perilous journey alone.

The Osprey is one of the most widely distributed species of bird on Earth. It is one of only six landbird species to occur on every continent except Antarctica during the course of the year, with many northern populations migrating south for the winter. The most northern Ospreys breed within the Arctic Circle in Finnish Lapland, while some Finnish birds have been found to migrate as far south as the southern tip of South Africa. In some places, the Osprey is referred to as a sea hawk or fish hawk, but in fact it is not a hawk at all, and instead is the only living species in the Pandionidae family. The scientific name for Osprey, *Pandion haliaetus*, is derived from the mythical Greek king of Athens, Pandion I, and the ancient Greek word *haliaietos*, *hals* meaning 'salt' or 'the sea' and *aetos* meaning 'eagle'. This is a reference to the fact that the Osprey is often, but not exclusively, a bird of coastal habitats. It is the only diurnal bird of prey (in other words, active during the day) that feeds exclusively on fish, and will readily hunt along coasts and estuaries as well as in inland lakes, ponds and rivers.

Above: Ospreys catch their prey in a range of different habitats – along rivers, in inland lakes or at the coast. They then take the fish off to a favourite perch in order to eat it.

The basics

A mid-sized bird of prey, the Osprey is larger than the Common Buzzard (*Buteo buteo*) but smaller than most eagles, with a wingspan of approximately 1.5m (5ft). Ospreys have relatively long, narrow wings, and fly with strong, steady wingbeats mixed with long glides. They hold their wings in a shallow 'M' shape, giving them a distinctive flight profile. When hunting, an Osprey will frequently hover over the water, staring intently downwards as it searches for a fish. Females are slightly larger than males, and have broader wings, a heavier bill and thicker legs. They may weigh up to 2kg (4.4lb), and tend to be around 14 per cent heavier than males, but this varies between races.

Adult birds of both sexes are uniform brown above and white below, with varying amounts of brown streaking on the breast and underwings. Females tend to be more heavily marked than males, and often have a pronounced breast-band. Both sexes have a white head with a distinctive brown eye-stripe that extends down the sides of the nape to join the hindneck.

Adult Ospreys have a bright yellow eye. Juveniles look similar to adults but have a darker amber eye and beautiful pale fringing to their brown upperparts. This fringing provides excellent camouflage in the nest and gives them a striking appearance when they fledge. Juveniles begin moulting their feathers once they arrive on the wintering grounds, and attain their full adult plumage before they are two years of age.

Above: Female Ospreys (right) are larger than males, with a more pronounced brown breast-band. Adult Ospreys have a bright yellow eye, but the male at the Cors Dyfi nest in Powys, Wales (left), known as Monty, has an amber eye, which is more typical of juveniles.

Vocalisations

The Osprey has a range of different calls, and it is possible to identify at least four different types of vocalisation. One of the most evocative sounds that can be heard in an Osprey colony during the breeding season is the high-pitched call of a displaying male. A distinctive *eep eep eep* betrays the presence of a male high in the sky, and he rises and falls as he utters the distinctive call, often hanging in the air with feet dangling before diving down and then rising up again, like a rollercoaster.

A short, sharp *chip* is uttered by breeding birds in response to the sight of an intruding Osprey approaching the nest, and this call increases in both frequency and intensity as the intruder gets closer. This same call is also given when two birds are hunting close together. Some Ospreys become highly possessive over specific hunting grounds, and will chase others away. Adult birds can be equally territorial during winter, and may be particularly aggressive towards newly arrived juveniles searching for somewhere to spend their first winter. If the initial warning *chip* is not heeded, they take flight and chase the juvenile away.

During the breeding season, male and female Ospreys have clearly defined roles, and females solicit for food throughout the summer by uttering a series of repetitive notes. This call will often prompt the male to go fishing, but on other occasions he will remain totally unresponsive, causing the female to beg even more loudly and persistently. Juveniles give the same food-begging call, particularly once they have fledged but are still dependent on their parents for food. The noisy food-begging of juveniles in late summer is a characteristic sound of Osprey nests worldwide. A final call sometimes heard at Osprey nests is a repeated *de de de* alarm call, given in response to people or a potential predator approaching the nest. This is similar to the food-begging call but shriller.

Above: An adult male Osprey (left) with one of his newly fledged offspring. Juvenile Ospreys have beautiful pale fringing to their brown feathers.

Race relations

Most taxonomists recognise four different races, or subspecies, of Osprey, categorised on subtle differences in size and plumage. *Pandion haliaetus haliaetus* occurs across much of Europe, including the UK, as well as north-west Africa and Asia north of the Himalayas. Within Europe, the Osprey's range extends from Lapland south to the Mediterranean islands, which is indicative of the species' ability to exploit a variety of different habitats for both breeding and foraging. Northern birds in Scandinavia and Russia tend to breed in forested areas, and hunt exclusively in inland lakes and rivers, whereas birds that breed in the Mediterranean region often build nests on sea cliffs and forage along the coasts. Despite this extensive geographical distribution, the European population is disjointed, and the Osprey remains absent from large parts of its former range due to historical persecution and the effects of harmful insecticides. Encouragingly, the species is now beginning to recover and is expanding across much of Europe, aided by translocation projects, which have been undertaken in six different countries (see pages 86–102).

Pandion haliaetus carolinensis breeds in North America, and is widely distributed across Canada and

Below: The European (left) and American (right) races of Osprey are very similar, but the American birds tend to be slightly larger and paler on the head and breast.

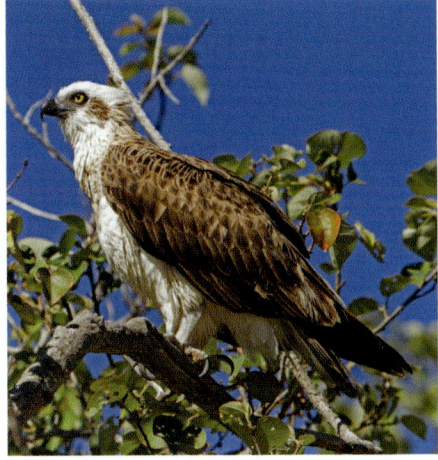

the United States. As in Europe, it is expanding in this region after suffering extensively from the effects of insecticides. There are now breeding records or nesting attempts for all of the lower 48 states of the United States, with core populations along the Gulf and Atlantic coasts. The highest densities occur on the Atlantic coast, with close to 10,000 pairs breeding around the Chesapeake Bay, a vast estuary in the states of Maryland and Virginia. In Canada, the range has expanded northward in recent decades, with pairs breeding as far as 68.3°N at Inuvik in the Northwest Territories. The American birds tend to be much more gregarious and approachable than their European counterparts, almost certainly as a result of a lack of human persecution. This sometimes results in the birds building nests in unusual locations, such as the masts of boats. *Pandion haliaetus ridgwayi* occurs in the Caribbean region, with the population distributed from the east coast of Yucatán in Mexico, south into Belize and east to the Bahamas, the Turks and Caicos Islands, and Cuba. *Pandion haliaetus cristatus* is found in the Indo-Pacific region and Oceania, with breeding populations in Indonesia, the Philippines, the Palau Islands, New Guinea, the Solomon Islands, New Caledonia and Australia. This race is sometimes recognised as a separate species, the Eastern Osprey (*Pandion cristatus*), although this has not been widely accepted.

Above: Caribbean Ospreys (left) have a distinctive white head with a reduced eye stripe compared with the other races, and are also very pale on the breast. Australasian birds (right) are smaller, with a limited eye stripe compared with European and North American Ospreys and extensive white on the head.

On the move

Northern populations of *Pandion haliaetus haliaetus* and *P. h. carolinensis* are migratory, with European and Asian Ospreys wintering in sub-Saharan Africa, the Indian subcontinent and South East Asia, respectively; and North American birds wintering in South America. In contrast, the Caribbean and Australasian birds are mostly sedentary, with some limited post-breeding movements. Similarly, individuals from lower latitudes such as Florida, the Atlantic islands and the Mediterranean may be sedentary, or make limited post-breeding movements.

Ospreys migrate alone, and without experienced adult birds to follow or the benefit of previous migratory experience, juvenile Ospreys must rely on an inherited programme of direction and distance, known as vector summation, to help them navigate to distant wintering grounds thousands of kilometres away. To make matters even more difficult, they must also learn to catch fish for themselves as they migrate south. Survival rates of juvenile Ospreys in migratory populations are consequently very low, with as few as 20–30 per cent of birds living to two years of age.

Below: In contrast to Ospreys from more northerly regions, those that breed in the Caribbean and Australasia do not need to migrate and instead tend to be sedentary throughout the year.

Most young Ospreys linger on the wintering grounds for the whole of their second summer, and head north again only once they are approaching two years of age. At this point they are drawn back to their natal site, and as they fly north, they often return to places they encountered on their first migration south. Male birds in particular are highly site faithful, and often return to breed just a few kilometres from where they fledged.

Threats to Ospreys

Over the centuries, Ospreys have greatly suffered at the hands of humans. They were relentlessly persecuted in Europe, beginning in the medieval period and continuing until the species was eradicated from many areas. Then, in the mid-20th century, the use of harmful agricultural insecticides had a further devastating impact on the distribution of the species in Europe and North America. Happily, the Osprey is now a protected species, and in recent years conservationists in both North America and Europe have helped restore the birds to areas where they had been lost through artificial nest-building and translocation projects (see pages 86–102). The species is expanding in many areas as a result, but threats still remain.

Above: Building artificial nests has helped Osprey populations to recover in both North America and Europe. This juvenile Osprey is perched on an artificial nest in Wisconsin, along the shores of Lake Michigan.

Diet and Feeding Behaviour

The one common trait shared by all Ospreys around the globe is their diet. As a specialist piscivore, the Osprey is supremely well adapted for catching fish in a variety of habitats from coasts to inland lakes and rivers. The sight of this awe-inspiring predator crashing into the water at high speed and then lifting off a few seconds later with a fish grasped in its huge talons, is a sight to behold.

The hunt

Ospreys will readily fish along coasts and in estuaries, or on inland lakes, rivers and even small ponds; the critical requirement is that fish are close to the surface. A hunting Osprey may use a variety of techniques but most commonly will circle or hover over the water at a height of 5–40m (15–120ft). If it is windy, the bird will always turn to face into the wind, holding its position with remarkable poise even in gusty conditions. Its pale underside helps it to remain inconspicuous against the bright sky, and a high density of eye receptors provides it with excellent clarity of vision.

Upon sight of a fish, the Osprey folds its wings and plummets towards the water, tracking the fish all the while. Often the bird will pull out of a dive at the last moment, not wanting to waste unnecessary energy if the fish dives out of reach or is too large. If its intended target remains within range, the bird flings its talons forward – as it prepares to strike, its feet are just millimetres in front of its eyes. Then, as it enters the water, a transparent third eyelid, or nictitating

Opposite: After a successful hunt the Osprey will always take the fish back to its nest-site or a favourite perch in order to eat it. Osprey may catch fish that can weigh more than half their body weight.

Below: A hunting Osprey will often hover as it searches for a fish.

Above: A diving Osprey throws its huge talons forward just before it hits the water.

membrane, closes for protection and the bird splays its huge feet.

The Osprey's lanky legs enable it to reach down to a depth of 1m (3ft), and its large scaly feet and their razor-sharp talons provide a tight hold on the slippery fish. Small spines – or spicules – on the underside of the toe pads enhance the grip further, and a reversible outer toe, which can face forwards or backwards, gives the bird extra manoeuvrability as it attempts to keep hold of the fish and then lift it out of the water. A sure sign that an Osprey has captured a fish is the sight of it resting on the water with wings outstretched as it ensures it has a firm grasp, before mustering sufficient energy to lift its catch out of the water.

Ospreys have oily feathers to avoid getting waterlogged while grappling with a fish, and this is vital because it can sometimes take several attempts for the bird to lift its catch out of the water. Fish caught typically weigh 150–300g (5–10oz), but some may be as heavy as 1kg (2.2lb), which is more than half the body weight of the average Osprey. Getting airborne with a large fish is no easy task, but a series of powerful, almost horizontal wingbeats will eventually enable the bird to take off.

Ospreys always carry their prey head first and, as the bird leaves the water, it is not uncommon for it to be gripping the fish with only one foot. Sometimes, this results in the Osprey dropping the fish, but usually it is able to gain a secure hold with one foot in front of the other so that the fish's head points forward. This improves aerodynamics, which is particularly important when carrying a heavy fish into a headwind. As the Osprey flies away, a characteristic shimmy in mid-air removes any excess water, and the bird heads off to its nest or favoured perch. The success rate varies according to a variety of factors, including weather conditions, water clarity and depth, and the intended prey species, but on average, about one in four dives results in a catch.

Below: Once it has grasped hold of a fish, an Osprey manoeuvres its catch so that it carries it off head forwards, in the most aerodynamic position.

On the menu

The Osprey's hunting habits reflect the fact that it can catch fish only on or close to the surface. As a result, river estuaries and coastlines, as well as shallow lakes, are particularly favoured hunting grounds, both in summer and winter. At the most northerly latitudes, Ospreys must delay their arrival at their breeding grounds in order to wait for lakes to thaw after the winter freeze. They are certainly not picky when it comes to selecting their prey.

Coastal catches

Flounder live on the muddy bottoms of oceans and estuaries but are often a favoured prey species of coastal Ospreys. You might think that they would be almost impossible to catch, but flounder inhabiting the shallow water of estuaries are exposed as the tide falls. They are often cut off from the main channel as the water recedes,

Below: Despite inhabiting the muddy bottom of estuaries, flounder are often caught by Ospreys who hunt for them at low tide.

and puffs of mud – which are created as they feed – betray their presence to hunting Ospreys, which know that a falling tide offers them the best chance of success. This means that despite their bottom-dwelling habits and inconspicuous nature, flounder are a staple in the diet of Ospreys in coastal regions of both North America and Europe.

Coastal Ospreys have also learnt to be opportunistic, taking advantage of seasonal variations in fish abundance, and changing their hunting habits accordingly. For instance, in Nova Scotia in Canada, Ospreys catch Pollock or Saithe (*Pollachius virens*) and Atlantic Herring (*Clupea harengus*) in spring, but once these fish disappear later in the summer, they resort to flounder. Other coastal and marine species regularly caught in North America and Europe include mullet, Sea Trout (*Salmo trutta*) and Atlantic Herring.

Freshwater favourites

Trout are a favoured catch of Ospreys fishing inland lakes in northern Europe, but a range of other species are taken, including Roach (*Rutilus rutilus*), European Perch (*Perca fluviatilis*), Pike (*Esox lucius*) and Common Bream (*Abramis brama*). Roach often occur in shoals near the surface of the water, making them relatively easy to catch. Bream, on the other hand, are bottom feeders, but Ospreys hunt them when they are spawning in shallow water. Such opportunism means that there is often some seasonal variation in the composition of an Osprey's diet, even in inland locations.

Varied diet

The migratory habits of Ospreys from northern latitudes means the composition of their diet, as well as the required fishing habits and techniques, may vary considerably between summer and winter. Ospreys that spend the summer catching trout in inland lakes in northern Europe may survive on a diet of saltwater fish during winter. Most Ospreys from northern and western Europe spend their winter on the coast of West Africa. At sites such as the Sine-Saloum Delta in Senegal – a hotspot for European Ospreys (see page 65) – adult birds fish in the shallow tidal waters once or twice a day. Shoals of two different species of mullet (*Mugil cephalus* and *M. curema*) make for easy pickings, but more exotic-looking species, such as Senegal Needlefish (*Strongylura senegalensis*), are also caught. Ospreys must be careful to avoid the sharp beaks of needlefish, which can impale them on impact, leading to serious injury.

Below: Ospreys in West Africa often catch needlefish.

Fishing trips

Above: Male Ospreys provide the majority of fish for their family during the breeding season.

Many Ospreys breed close to favoured fishing locations. Some are able to catch fish within sight of the nest, like the well-known Manton Bay Ospreys at Rutland Water in central England, and even from favourite nestside perches. Others fly considerably further. In recent years, satellite tracking has shed new light on the foraging behaviour of individual Ospreys, and demonstrated that some birds are prepared to travel surprising distances to favoured fishing grounds.

Male Ospreys undertake most of the fishing during the breeding season and, at the peak of activity, when there are small chicks to feed, a breeding male will catch his first fish at first light and then make regular foraging trips throughout the day. On some days he will deliver as many as five or six fish, often caught from many different sites, to his mate and offspring. Satellite-tagged males in northern Scotland alternate between fishing spots on the coast and inland lochs, while at Rutland Water some breeding males regularly make round trips of more than 40km (25 miles) to favourite sites. This means that a male Osprey will travel thousands of kilometres during the course of a breeding season in order to feed his family, building up an extensive knowledge of all waterbodies in the area.

Success rates

The duration of foraging trips varies according to many different factors, most obviously the distance between the nest-site and fishing grounds, but also a range of environmental variables. Approximately one in four dives by adult Ospreys is successful, but weather conditions play a crucial role in influencing this strike rate. High winds and heavy rain can be especially problematic, even for experienced adult Ospreys. Gusty conditions make it difficult for a fishing Osprey to maintain its position above the water, and also create waves on the surface that make it more difficult to spot fish. In such conditions, Ospreys will usually favour sheltered places where they are protected from the strongest wind. Heavy rain may preclude fishing altogether and sometimes force a hunting Osprey to abandon fishing temporarily while it waits for the rain to cease.

Below: In North America, Bald Eagles (right) will often attempt to steal fish from Ospreys.

At the opposite end of the spectrum, bright sun can create glare on the surface of the water that also makes it difficult to locate fish. This means that days with little or no wind and some cloud are perfect. If weather conditions are not in their favour, male Ospreys can be away from their nest for several hours at a time and may make numerous unsuccessful dives before finally making a catch.

Once they have seized a fish, Ospreys are often harassed by large gulls that sometimes force them to drop their catch. In North America, Bald Eagles (*Haliaeetus leucocephalus*) may also attempt to rob Ospreys of their prey, as may White-tailed Eagles (*H. albicilla*) in Europe, or African Fish Eagles (*H. vocifer*) on the wintering grounds. On occasion, an adult male Osprey will eat a fish away from the nest, before returning to catch a second fish to take to his waiting family.

Favoured sites

Fishing Ospreys can be surprisingly bold, with some individuals allowing a very close approach, and often diving for fish near boats. In some places, they have learnt to exploit fisheries and fish farms, where fish are kept in high densities and are very easy to catch. At Rothiemurchus Fishery in northern Scotland, and River Gwash Trout Farm near Rutland Water in central England, this problem has been turned around through the

Below: Ospreys have learnt to exploit fish farms in some areas, but any potential conflict has been avoided in some places through the construction of hides that provide an opportunity for wildlife photographers to get close to fishing Ospreys.

Above: Ospreys sometimes catch two fish in a single dive if they target a shoal of fish.

construction of specially designed hides that offer wildlife photographers a unique opportunity to capture the birds in action. The hides have proved extremely popular and have generated a valuable income in both cases, while the stunning photographs speak for themselves.

Adult Ospreys return to the same wintering site each year and, as in summer, they become familiar with the best fishing locations. The distances wintering Ospreys fly during foraging trips vary by location, but satellite tracking has demonstrated that some birds have minimal winter ranges, often as little as a few square kilometres. For example, birds that winter along the coasts of Senegal, the Gambia and Guinea-Bissau will usually make one or two flights out to sea to catch fish each day, and spend the rest of the time perched on beaches or in coastal mangroves. Many foraging flights are just a few hundred metres offshore, but occasionally Ospreys will fly much further out to sea to take advantage of shoals of fish. In such a situation, an Osprey will sometimes pull more than one fish out of the water in a single dive, although this is likely an unintended benefit of targeting a shoal of fish.

Breeding

Ospreys remain faithful to the same mate and nest-site year on year. After spending the winter apart, the pair is reunited when the birds return to their large stick-built eyries each spring. Male and female Ospreys have clearly defined roles during the breeding season, with the female undertaking most of the incubation and then all of the subsequent brooding and protection of her young. The males, meanwhile, must defend the nest and provide a constant supply of fish to satisfy the needs of his hungry family. Osprey chicks fledge at around seven weeks old and embark on their first migration a few weeks later.

Osprey nests

The timing of arrival of Ospreys at their nests varies according to latitude, with some Mediterranean birds back at nests in late February, but the most northern birds – such as those that breed in Finnish Lapland – delaying their arrival until well into May. Ospreys build conspicuous nests on the tops of trees or other structures, and the best sites may be used for many decades by successive generations, some becoming enormous as a result. They like a clear, unobstructed view, and

Opposite: In Scotland and other northern countries, Ospreys usually build their nests in the tops of prominent pine trees.

Below: Ospreys will sometimes nest on very low trees, particularly in open areas such as in saltmarshes and river estuaries. Their main requirement is a clear, unobstructed view.

Above: Ospreys readily use artificial nests constructed on top of telegraph poles, and become habituated to farming operations nearby.

Below: In some parts of the world, including North America, Ospreys build their nests on a range of man-made structures. Some nests are used for many generations and can become huge as a result.

usually choose an isolated tree, or the most prominent tree in a small group, on which to construct the nest, which is built with sticks and then lined with turf and moss. Many Scottish nests sit atop very tall Douglas Firs (*Pseudotsuga menziesii*) topping 30m (100ft) or more, providing a commanding view of the landscape but leaving them exposed to the elements, particularly during winter, when nests may collapse after heavy snow.

In other parts of the world, all manner of structures are used. In Germany, more than 400 Osprey pairs breed on the top of electricity pylons, while in the United States nests are built on TV aerials, channel markers in estuaries and even atop the masts of boats. Mobile phone masts have also been used in several countries, including Scotland. In Corsica, breeding Ospreys nest on immense sea cliffs, while in some parts of the world they even nest on the ground. This has been recorded twice in Scotland, but sadly the nest was predated on each occasion.

Ospreys will readily use artificial nests built in trees or on top of telegraph poles, which has been an excellent way to encourage the species to spread to new areas (see pages 79–85).

Housekeeping

When Ospreys return to their nest in the spring, they usually spend several days repairing any wind damage. Some of the larger, more robust nests suffer very little over the winter, whereas those in more exposed sites require several days of repair before they are ready for the new breeding season.

Both the male and female build nests, but the male tends to do the bulk of the work in the spring, breaking branches from nearby trees or swooping to pick up sticks from the ground. Some of the sticks measure 30–40cm (12–16in) in length but others are much longer; it can be quite a sight watching a male Osprey flying to the nest with a metre-long (3ft) stick trailing behind him. Back at the nest, the sticks are meticulously moved around and interwoven to form a remarkably secure structure. Some Ospreys continue adding to it throughout the summer, meaning that by the end of the breeding season the nest will have grown in size considerably. Others add one or two sticks when they first arrive in the spring but very few after that. Such differences appear as much to do with the character of individual birds as with the site they occupy.

Above: Ospreys construct their huge nests with sticks that they break from trees or pick up from the ground.

Another task when an Osprey returns to its nest is to scrape out the nest cup. By the end of the breeding season, the middle of the nest becomes flattened, thereby creating a firm base for the young to launch themselves into the air for the first time. This, however, is no good for protecting the eggs, and so upon arrival in spring, the breeding pair create a shallow depression in the centre of the nest by leaning forward and then kicking backwards with their powerful feet, sending a shower of old lining behind them in the process. This depression is then lined with clumps of fresh moss and turf – and even the odd dry cowpat in grazing areas – brought in by both the male and female in preparation for incubation. In fact, a sure sign that a female is preparing to lay the first egg is an increase in the number of clumps of moss and turf that she brings to the nest.

Below: Both male and female Ospreys participate in nest-building, although males often do the majority of the work in spring.

Reunited in spring

Above: Egyptian Geese have become a competitor for nests in some parts of the world. The resident Ospreys chase them away from established sites.

Most Ospreys remain faithful to the same nest-site and mate, and often become very predictable in their arrival date, returning on or very close to the same date each year. Long-standing Osprey nests are highly sought after, particularly in established colonies, and so a timely arrival in spring ensures that the regular nest-holders do not get usurped by rival birds. And it is not just other Ospreys that they must be wary of; Canada (*Branta canadensis*), Greylag (*Anser anser*) and Egyptian (*Alopochen aegyptiaca*) geese all compete with Ospreys for their nests, as do White Storks (*Ciconia ciconia*) in some areas. If any of these species get established on a nest early enough in the spring, Ospreys find it difficult – if not impossible – to oust them.

In the UK, one of the first Ospreys to arrive each spring was 03(97) – or Mr Rutland, as he became better known. This male, which was translocated from northern Scotland to Rutland Water in 1997 as part of the first European reintroduction project, bred for 15 summers at a nest he built at the top of an oak tree near the reservoir in the summer of 2000. He raised a single chick with an unringed female for the first time in 2001, and went on to rear a total of 32 chicks with three different partners. During this period, he was usually the first Osprey to return to Rutland – and in some cases the UK – sometimes as early as 15 March.

It is always exciting to watch a newly arrived Osprey back at its nest in the spring, and even more so if you are lucky enough to witness the moment a pair of Ospreys reunite after a winter apart. The male usually greets the arrival of his mate with a spectacular undulating display flight, rising and falling in the air while giving a distinctive and evocative *eep eep eep* call. However, it is not always the male that gets there first. Older females sometimes arrive before their mate, particularly if they winter at a more northerly latitude. There is an increasing trend for northern European Ospreys to winter in Iberia, and there is no doubt that these individuals are at a distinct advantage in the spring compared with birds migrating north from sub-Saharan Africa.

Below: Translocated Osprey, 03(97), bred for 15 years at a nest close to Rutland Water and was often the first Osprey to return each spring.

Defending the nest

Unlike raptors such as Golden Eagles (*Aquila chrysaetos*), which defend a huge territory centred on their nest, Ospreys are semi-colonial breeders, meaning that they actually prefer to nest close to other Ospreys. In some parts of the world, nests are incredibly close together. In Chesapeake Bay on the east coast of the United States, for example, where around 10,000 pairs of Ospreys breed on saltmarshes and surrounding areas, nests can be as close as 30m (100ft) to one another. In Florida, meanwhile, there are some sites where two pairs of Ospreys will actually breed in the same tree. Even in Scotland, where the population density is considerably lower, there have been active nests just a few hundred metres apart, although it is more common for nests to be separated by 2–3km (1.2–1.8 miles) or more.

During the breeding season, male Ospreys defend their nest against other Ospreys, uttering a high-pitched *chip* if a rival bird approaches too close, and then giving chase if the intruder does not heed the warning. Young birds attempting to breed for the first time prefer to take over an established site rather than build their own, and this often leads to fights over long-standing nests, particularly if regular breeders arrive late in the spring (see page 32).

Above: Ospreys are a semi-colonial breeder and like to nest close to other Ospreys. In some parts of the world, most notably in North America, nests can be just a few metres apart in the same tree.

Young usurpers

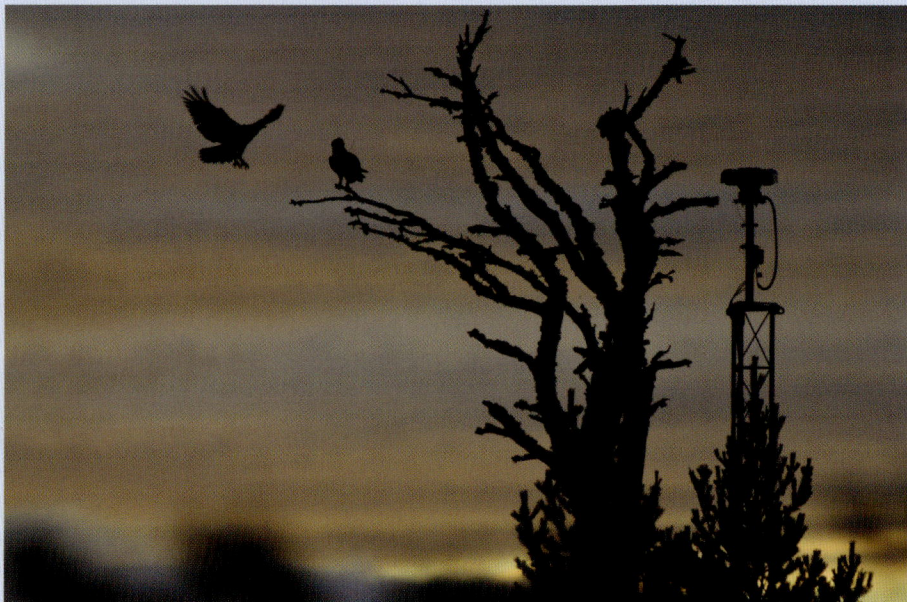

Above: Ospreys at the famous Loch Garten nest in northern Scotland were embroiled in an amazing saga for three years running between 2005 and 2007.

The issues that can arise if an established breeder arrives late in the spring were exemplified at the famous Loch Garten nest in northern Scotland between 2005 and 2007, in a saga that made national headlines. In 2004, a male known as Henry and a female called EJ reared three chicks. The next spring, EJ was back on 29 March and, in the absence of her mate, paired with a rival male, Red 8T. Henry eventually arrived on 28 April in poor condition, suggesting he had undergone an arduous migration. Nevertheless, he ousted Red 8T and scraped out the nest cup, kicking out four eggs laid by EJ. It was now too late for her to relay, but the birds spent the summer together, allowing Henry to recover from his difficult migration.

In the spring of 2006, EJ returned on 26 March and initially paired with Orange VS, a male she had bred with at another nest from 2001 to 2003. Fortunately, on this occasion, Henry arrived on 10 April, before EJ had laid eggs. After three hours of fighting, he successfully reclaimed the nest, and he and EJ went on to raise three chicks.

In spring the following year, EJ arrived later, on 4 April, but again before Henry. As in the previous spring, she paired with Orange VS, who had been present at his own nest since 30 March. Over the next two weeks he travelled between the two sites, mating with EJ and providing sporadic fish deliveries. When Henry finally returned on 22 April, EJ had already laid two eggs. Henry immediately removed them and, when EJ laid a further two eggs 48 hours later, he did so again, apparently sensing that they were not his. Henry and EJ continued to mate for the next two weeks, and EJ finally began laying a replacement clutch on 11 May, allowing incubation to start in earnest. Sadly, two chicks died 24 hours after hatching, and the third perished as it attempted to break out of its shell, meaning that for the second time in three years the late arrival of Henry had resulted in him and his mate failing to raise any chicks.

Multiple mates

An excess of unpaired males often leads to fighting over established nests, but if the sex ratio of the population is skewed towards females, then a different situation may play out. Polygyny – where a male breeds simultaneously with two different females – is relatively rare in Ospreys but has been recorded at a number of locations around the world. Alan Poole, a world-renowned expert on American Ospreys, recorded three polygynous trios out of 190 monogamous pairs he monitored on the east coast of the United States in the early 1980s.

More recently, polygyny has been recorded in the expanding Welsh population and also at Rutland Water. In each case, an unattached female has settled on a vacant nest within close proximity (usually less than 2km, or 1.2 miles) of an established pair. If she remains unpaired, the female will accept the male from the rival nest, and mating and egg-laying inevitably follows. In the early stages of incubation, the male often divides his time between the two nests, providing fish to both females and taking his turn incubating at both sites. However, as the days and weeks progress he usually begins favouring his established site, forcing the second female to begin fishing for herself, and leaving the eggs unattended in the process. As a result, it is extremely rare for the second female in a polygynous trio to produce young, and it has never been documented in the UK.

Below: Very occasionally a male Osprey will pair up with females at two different nests. It is very rare, however, for both nests to be successful.

Above: Male and female Ospreys have clearly defined roles during the breeding season. Females undertake the main bulk of incubation with the male providing a regular supply of fish.

Right: When a male Osprey returns to the nest with a fish he has usually eaten the head first before giving the remains to his mate.

vociferous. Eventually, the male will take his cue and set off in search of a meal.

The female is more vulnerable while her mate is away from the nest, and intrusions by non-breeders can be problematic when the male is off fishing. In the expanding colony at Rutland Water, there have been instances of a reduced brood size, or nests failing altogether, as a result of persistent intrusions by young non-breeding males. Although Osprey eggs are fairly robust, they can become chilled if the female is kept off the nest by an intruding bird. Leaving the eggs unattended in order to chase off an intruder also increases the risk of an opportunistic crow or other corvid predating them.

When the male does eventually return with a catch, he will usually eat the head of the fish first before delivering the remainder to his waiting mate. This signals a change in incubation duties, with the male settling down on the eggs and the female taking the fish to a favoured perch in order to feed. That said, it is not always a fish delivery that signals a swap-over, as the male takes his turn at sitting at regular intervals during the day. Sometimes, this can be for just a few minutes, while at other times it may last for an hour or more. The female, though, generally undertakes the night shift.

Osprey nestlings

Newly hatched chicks

Osprey chicks are extremely weak at first, but within a few hours of hatching, their neck is strong enough for them to lift their head up and beg for food. Even at this very young age they already resemble their parents, with a distinctive brown eye-stripe on their white downy head.

The sight of a newly hatched chick prompts a distinct step change in the frequency of the male's fishing trips, and he immediately doubles or even triples the number of fish he delivers to the nest each day. As happens during incubation, he will often eat the head of the fish before taking the remainder to the nest, but now, rather than flying off to eat the fish, the female remains on the nest. She takes the fish from the male and then slowly inches her way to the side of the nest, where she tears off tiny morsels and delicately offers them to her newly hatched chick. The chick will readily take food within a few hours of hatching, but only in very short stints of between five and 10 minutes. Once the chick is full, the female will leave the fish on the side of the nest and settle back down on the chicks and any remaining unhatched eggs. The hatching of the first chick heralds a change in the roles of the adult birds: the male will now concentrate solely on catching fish and protecting the nest, leaving the female to brood and care for the young.

Above: Osprey chicks hatch after around 37–41 days of incubation, and in the order the eggs were laid. Newly hatched chicks are brooded by their mother to keep them warm.

Above: Once the male Osprey has delivered a fish to the nest the female takes it from him and offers tiny morsels to the chicks.

By the time the second and third eggs hatch, the oldest chick will have gained weight and, as a result, is dominant at feeding time. However, as long as the male provides a plentiful supply of fish, the female will continue to feed any weaker and subordinate siblings once the largest of the brood is full. Weather at hatching is crucial to the survival of the young. Cold and wet conditions can be fatal, through the combined effect of chilling the chicks and making fishing difficult for the male. In such a situation, the youngest of the brood invariably do not survive. However, female Ospreys are incredibly strong and determined mothers, sheltering their young remarkably well from torrential rain and cold winds. This is particularly important given the Osprey's preference for exposed nests. A nest on top of a 40m (130ft) Douglas Fir in northern Scotland gets more than its fair share of inclement weather in a typical summer.

The reptilian stage

Osprey chicks grow exceptionally quickly thanks to their protein-rich diet. Within 10–12 days of hatching, they enter the 'reptilian stage', when the emergence of their second down gives them a much darker, scaly appearance and they resemble miniature velociraptors. At this stage, the youngsters' appearance changes markedly by the day, but there is often still a size difference, particularly between the oldest and youngest of the offspring. Squabbles between siblings sometimes occur – particularly as the chicks become more mobile – but rarely amount to anything serious unless there is a food shortage, in which case any size difference may become more pronounced as the eldest dominates at feeding time.

The first rusty-brown pin feathers appear on the chicks' head and neck once they are a fortnight old, and body feathers follow soon afterwards. By three weeks of age, the chicks are extremely mobile but are still not strong enough to walk on their feet. Instead, they shuffle around the nest on their haunches with their disproportionately large feet giving them a decidedly comical appearance. They often manoeuvre their way to the edge of the nest and stare at the world below. By now, their wings are starting to develop, with primary and secondary feathers beginning to grow. The chicks are so large that it is impossible for their mother to brood them, but she does her best to protect them from heavy rain and also strong sunshine.

During the early stages of the nestling period, a male Osprey with a clutch of three hungry chicks will deliver as many as five or six fish to the nest every day, and, as a consequence, it is essential that he maximises all available daylight by fishing at first light and then at regular intervals until dusk. Whenever he lands with a fish, the chicks, which are often scattered around the nest, stand up and as quickly as their awkward movements will allow, make their way to their mother to be fed. By this stage, some males will offer one or two pieces of fish to their offspring themselves, but most leave this critical role to their mate.

Above: Osprey chicks enter the 'reptilian stage' once they are 10–12 days old. During this period they resemble miniature velociraptors.

Bloom of youth

By the time they are a month old, the chicks will have attained around three-quarters of their body weight and will be fully feathered, albeit with some growing of primaries and secondaries still to do. Beautiful pale fringing to their brown upperparts gives the nestlings a

Above: By the time they are three to four weeks old, Osprey chicks have lost much of their down and are well feathered.

striking appearance, but crucially also offers camouflage. At the first sign of danger – usually signalled by an intruder or alarm call from their mother – the chicks immediately lie prone in the nest, their mottled plumage making them remarkably inconspicuous from above. This is particularly important in parts of the world where eagle owls and Bald Eagles regularly predate Osprey nestlings. Buzzards will also take Osprey chicks up to a surprisingly large size; nest-cam footage from Scotland recorded the moment a Common Buzzard swooped in and carried off a three-and-a-half-week-old chick with astonishing ease.

At four to five weeks of age, the youngsters are strong enough to stand up, and this enables them to start flapping their wings for the first time. Although somewhat weak and feeble at first, this wing flapping grows progressively stronger each day, and by the time the chicks are six weeks of age each stroke of their almost fully formed wings is considerably more powerful and controlled.

With her offspring now almost fully grown, there is much less space on the nest for the female, and so she usually sits on a nearby perch keeping a watchful eye over her brood. Another critical difference once the youngsters reach six to seven weeks of age is that they

are able to feed themselves. This means that at this stage when the male lands on the nest with his latest offering, there is often a clamour to take the fish from him. If the female gets there first, she will feed the chicks, but sometimes one of the youngsters will grab the fish, mantle over it with wings drooped and then expertly tear pieces from the flesh while holding the slippery skin in its talons. That said, it is rare for a chick to eat a whole fish in a single sitting, so as long as the fish is large enough, there is usually ample to go around.

Above: Osprey chicks grow incredibly quickly due to their protein-rich diet. At the height of the breeding season the male may deliver five or six fish to the nest each day.

Below: Osprey chicks are fully feathered by the time they are around five-and-a half weeks of age, although their flight feathers are still growing at this stage.

By now, it is possible to sex males and females even from a distance. Females are larger, chunkier birds with slightly broader bills and thicker legs. Males, in comparison, look relatively slim and sleek, although there are always individuals that lie somewhere in between and, as a result, are more difficult to separate.

Lift-off

Most Ospreys fledge between seven and eight weeks of age. During the week leading up to their maiden flight, their wing flapping becomes so powerful that it lifts the youngsters clear of the nest – usually by only a few centimetres at first, but after three or four days of this 'helicoptering', they sometimes hover a metre (3ft) or more above the nest. This is a sure sign that the juvenile Ospreys are ready to fly, and the only barrier left facing them isn't a physical one. At this point, it is merely a case of them mustering the courage to launch themselves from the safety of the nest for the first time. Sometimes this will happen entirely by chance when a rogue gust of wind while the juvenile is helicoptering results in a premature or at least unexpected first flight for them. On other occasions, a juvenile will suddenly leap into the air from the edge of the nest. No matter what the exact mechanism of fledging is, the one guarantee is that the chicks' mother will be on high alert, intently watching her offspring's every move at this perilous time. If one of her brood takes to the air, she will often fly with it, shadowing the youngster closely during its first experience of life in flight, and ensuring that it doesn't attempt to fly too far.

Above: Osprey chicks often 'helicopter' above the nest for several days before finally making their first flight.

Once airborne, most young Ospreys fly amazingly skilfully, but landing often takes them longer to perfect. In many ways the most dangerous aspect of fledging isn't learning to fly, it's learning to land safely again. Returning to the nest is the safest option after a maiden flight because landing in nearby trees or other perches takes some time to master. This is particularly the case during gusty conditions or heavy rain, when the chances of missing a landing and ending up on the ground increase significantly.

In the air

The post-fledging period – that is, the period between fledging and migration – is an incredibly important time for young Ospreys. It is an opportunity to gain essential life skills and to ensure that they depart on migration in good condition. In the first week after fledging, most young Ospreys do not venture far from the safety of the nest. They soon grow in confidence, making numerous short flights between favoured perches, but tend not to fly more than a few hundred metres at a time.

If you watch an Osprey nest at this time, you get a sense of the sheer enjoyment the youngsters gain during their first few days on the wing. If conditions are favourable, it is remarkable how quickly they adapt to life in the air. Within a few hours of making their first flight, they are able to perform the most incredible aerial acrobatics, twisting and turning in the air with aplomb and often engaging in aerial chases with other birds, such as corvids. Some Ospreys fledge at around seven weeks, but others take longer and, as a result, there is often a considerable time gap before all of the brood are flying. The youngsters yet to fledge watch on with great interest as their siblings perform flying circuits around the nest.

Above: Landing is often the most difficult aspect of a young Osprey's first few days on the wing.

Most young Ospreys start to venture further afield a week or 10 days after fledging. Initially, these exploratory flights are relatively short, but as the days and weeks progress the birds begin to wander further from the nest and can be absent for several hours. This behaviour is crucial, because it enables the juveniles to imprint on their natal area. Male Ospreys in particular usually return to breed close to the site where they were reared, and exploratory flights after fledging appear to play a crucial role in this. As the youngsters circle and soar over the

landscape, they are mapping all that lies beneath, learning that it is home. Satellite-tracking studies have shown that most exploratory flights are within a few kilometres of the nest, but the youngsters sometimes range much further – perhaps as far as 15–20km (9–12 miles).

While exploratory flights are important, being absent from the nest can have its disadvantages. Juvenile Ospreys remain dependent on their parents for food until they depart on migration, and a crucial element of the post-fledging period for any young Osprey is to ensure that it feeds well before it migrates. Setting off underweight and in poor condition is potentially fatal. Adult males continue to provide for their offspring until they set off on migration, but the trouble for any young Osprey with siblings is that there is intense competition for food. If one is absent from the nest when a fish is delivered, then the chances are that it misses out on a meal. In the nestling stage, the youngsters are fed by their mother, but by the time they are flying they are more than capable of feeding themselves. There is a great squabble each time the male lands with a fish, and the juveniles soon learn to carry food away to a favourite perch. By now, the youngsters have found their voices and their monotonous food-begging calls fill the air when they are anywhere near the nest. The instinct to feed well before migration is so overwhelming, young Ospreys will often give this call even when they are holding onto a fish.

Above: Juvenile Ospreys venture progressively further from the nest during the post-fledging phase, sometimes up to 20km (12 miles) away.

Final preparations

Once her chicks have been flying for several weeks, the adult female will often take the opportunity to depart on migration, leaving her mate to continue to provide for the young. During the latter stages of the post-fledging period, the male will often be hounded by hungry juveniles, which perch next to him and beg for food

incessantly. It is no surprise, therefore, that many males are absent from the nest for prolonged periods at this stage. They will often eat fish away from the nest, before eventually returning with a second catch to deliver to their offspring. This pattern continues until the last of the juveniles has set off south, at which point the male takes his cue to leave.

Although most juvenile Ospreys do not catch their own fish before they set out on migration, they do make practice dives into the water. These dives are extremely ungainly at first, but they improve as the days and weeks progress. Nevertheless, since the male continues to provide fish until the youngsters depart, there is little incentive for the juveniles to persist until they catch a fish. Occasionally, one will be successful, but this is very much the exception to the rule.

Above: It is rare for a juvenile Osprey to catch a fish before setting out on migration, but they often make practice dives into the water.

Left: Juvenile Ospreys finally begin to fish for themselves once they have set out on their first migration.

Migration

The cosmopolitan distribution of the Osprey means that migratory populations are found at northerly latitudes across much of the globe. Ospreys in Europe, Asia and North America are almost exclusively migratory, with adult and juvenile birds leaving northern breeding grounds in late August and September. Juvenile Ospreys migrate alone, driven by instinct and at the mercy of the weather encountered en route. In recent years, satellite tracking has revealed remarkable feats of endurance and helped us understand how Ospreys navigate and learn from experience during migration. Adult Ospreys return to the same wintering site each year, but for juvenile birds, finding somewhere safe to spend the winter is yet another tough challenge they must face.

Ospreys from western Europe migrate south along the East Atlantic Flyway – a route used by millions of migrating birds between northern breeding grounds and wintering sites in western Europe and along the western seaboard of Africa – with birds from the UK, France, Germany and Sweden travelling south through Spain. An increasing number of birds winter on the Iberian Peninsula, but most continue south, crossing the Strait of Gibraltar to Morocco, and then skirting around the western edge of the Sahara towards the

Opposite: All northern populations of Ospreys are migratory, with birds travelling to distant wintering grounds across some of the most inhospitable terrain on the planet.

Below: Unlike many migratory raptors, both adult and juvenile Ospreys migrate alone.

fish-rich coastal waters of West Africa, from Mauritania south through Senegal, the Gambia and Guinea-Bissau, towards the Ivory Coast and Ghana. Ospreys from populations further east, including Finland, eastern Europe and western Russia, tend to use a more easterly route that takes them south through the Balkans and the Middle East, and then south along the Great Rift Valley towards wintering grounds in eastern and southern Africa. Some Finnish birds, however, use a route more typical of their western neighbours and winter in West Africa. Meanwhile, those breeding further east in Russia migrate south to coastal Arabia, the Indian subcontinent and South East Asia.

In North America, migration routes are similarly divergent. Ospreys that breed in eastern parts of the United States and Canada tend to migrate south along the east coast before crossing from Florida to Cuba, and then continuing onwards through Haiti and the Dominican Republic, before crossing the Caribbean Sea to Colombia or Venezuela. Some birds winter in these northern parts of South America, while others continue south to winter in the Amazon Basin in Brazil. In contrast, Ospreys that breed in western parts of North America use an alternative route that takes them south through Mexico and Central America towards Peru, with some individuals wintering as far south as south-central Chile.

Below: Ospreys have a cosmopolitan distribution and occur on every continent except Antarctica at some point in the year. Most northern populations are migratory, whereas those from southern regions are sedentary or only make limited post-breeding movements. Yellow shaded areas show the summer (breeding) range, blue indicates the wintering range and green shows areas where ospreys are resident throughout the year. The red arrows on the map indicate some of the key Osprey migration flyways.

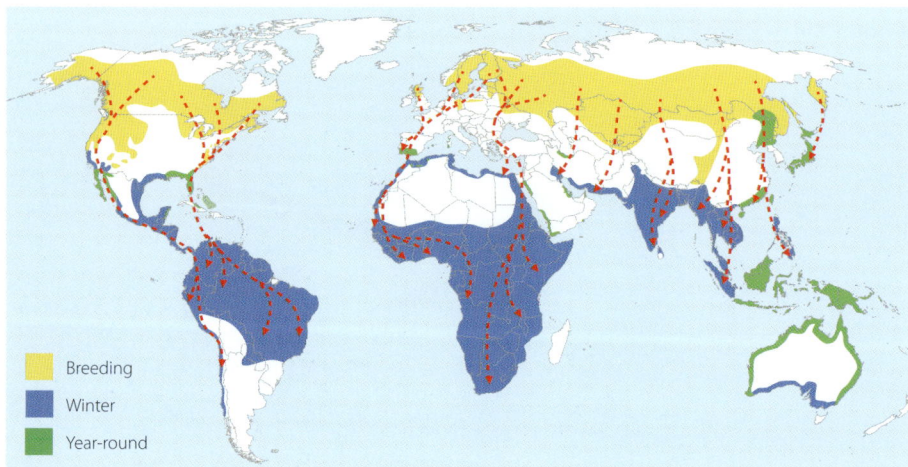

- Breeding
- Winter
- Year-round

Navigation

Above: White Storks migrate in mixed-age flocks, enabling juveniles to learn from experienced adults. Young Ospreys, however, rely on instinct alone to navigate to their wintering grounds.

Migration is one of the great wonders of the natural world, and the question of how birds navigate has fascinated scientists since the first British ringed Barn Swallow (*Hirundo rustica*) was recovered in South Africa more than a century ago, a finding that provoked understandable astonishment at the time. Research has shown that, unlike many species, Ospreys migrate alone, making a juvenile's first migration a challenge not only of endurance, but also of navigation. Many species of migratory birds, such as cranes and swans, travel in family groups. Others, including Honey Buzzards (*Pernis apivorus*) and White Storks, migrate in mixed-age flocks, with juveniles following more experienced adults. In each case, social learning plays a crucial role in shaping a juvenile's first migration. For young Ospreys, however, the initial decision to set off, the route they take and their final wintering destination are all driven by instinct alone.

Researchers have proposed that a process known as vector summation helps juvenile Ospreys to navigate to distant wintering grounds thousands of kilometres away. Under this mechanism, first described in the late 1970s, the migratory journey is broken down into a series of flight steps – or vectors – that involve the bird flying in an inherited direction. In the case of UK Ospreys, this is south-west. Any subsequent changes in orientation are caused by environmental factors, such as wind. This means that under optimum conditions juvenile Ospreys from the UK migrate either to Spain or Portugal, or continue across the Sahara to West Africa.

Winds of change

The wind has a profound effect on bird migration, and juvenile Ospreys are likely to be blown off course as they migrate south for the first time. The UK's proximity to the western Atlantic Ocean means that strong easterly winds are particularly troublesome for young Ospreys, and often result in inexperienced birds making very long flights across the sea. As we will discover later in this chapter, the Osprey's morphology allows it to complete very long sea crossings, but such journeys can be especially dangerous for inexperienced juveniles.

In the late 1990s satellite tracking was still in its infancy, but for the first time, it was possible to plot the exact routes of individual Ospreys as they migrated south. This was incredibly exciting, but it soon revealed the realities of migration for young birds. In early September 2000, a young female Osprey known as T09, which had been released at Rutland Water as part of an exciting new translocation project (see page 86), set off on her first migration south, equipped with a battery-powered satellite transmitter. Twenty-four hours later, she was just off the Brittany coast, heading south-west across the Bay of Biscay. Sadly, she just failed to make the Spanish coast, and for the next two weeks signals were received that indicated her body was floating on the sea, before it eventually sank.

Anecdotal evidence suggests that exhausted Ospreys have sometimes been seen arriving on the north coast of Spain only to be mobbed and forced down by gulls – perhaps that is what happened to T09? Unfortunately, she was not the only young Osprey from Rutland Water to drown at sea. A year later, a juvenile female known as U04 also crossed the Bay of Biscay. She missed the north coast of Spain and was eventually found washed up on a beach on Fuerteventura in the Canary Islands, a further 1,700km (1,050 miles) south. These fateful journeys identified the Bay of Biscay as a hazard for young Ospreys, and subsequent satellite-tracking studies have confirmed this.

Above: Satellite tracking has revolutionised our understanding of Osprey migration in recent decades.

Biscay boost

Satellite tracking has shown that in some cases, the Bay-of-Biscay crossings of UK Ospreys are constituent parts of very long, non-stop flights at the beginning of their autumn migration. The best example of this was the autumn migration of an adult male Osprey known as Blue XD. This breeding male was fitted with a Global System for Mobile communications (GSM) transmitter at his nest in Strathspey in the Scottish Higlands by leading Osprey conservationist Roy Dennis in summer 2013. GSM transmitters, which send data through the mobile phone network, log the bird's position as regularly as once a minute. They enable the bird's flight path to be plotted in incredible detail, but also allow inferences to be made about how the bird is flying, thanks to the inbuilt altimeter.

Having left his nest-site early on 7 September, Blue XD migrated south through western England, reaching the Dorset coast by dusk. With clear skies and northerly tailwinds providing perfect migration conditions, he continued south over the English Channel, across the Brittany peninsula and then onwards over the Bay of Biscay. By dawn the next morning, he was halfway to northern Spain, and he eventually came ashore in Cantabria shortly before 11 o'clock that morning.

Below: Many Ospreys migrate south along the Atlantic coast of France, stopping off at estuaries to feed. However, when winds are favourable, adult birds will sometimes fly direct across the Bay of Biscay.

Most Ospreys maintain a fairly constant altitude while migrating over the sea, indicating that, as expected, they travel predominantly using flapping flight. However, close examination of the GSM data showed that during the final 50km (30 miles) of Blue XD's Bay-of-Biscay crossing, his flight profile was more akin to flying over land than over sea, with altitude gains of 250–300m (820–980ft) in periods of five to six minutes, followed by long descents. Subsequent analysis of later migration tracks by the same Osprey, as well as those of two other adult male Ospreys from Scotland, revealed similar flight profiles during a quarter of ocean crossings. It is possible that the birds are able to achieve soaring and gliding flight over the sea by exploiting weak sea thermals that occasionally form under specific environmental conditions, or by skillfully using elements of the wind to help them gain altitude. Either way, travelling in this manner saves them valuable energy during long ocean flights.

When he was back over land, Blue XD showed no signs of letting up. He eventually settled to roost beside a river in Extremadura in central western Spain after an incredible non-stop flight of more than 2,000km (1,240 miles) in 36 hours. Satellite tracking helped us to understand that, at the beginning of their migration from the UK, Ospreys often follow Blue XD's strategy of extending their flight into the hours of darkness to take advantage of supportive tailwinds and clear, moonlit nights.

Below: Satellite tracking has shown that some experienced adult Ospreys take advantage of tailwinds at the start of migration by making very long continuous flights from Scotland direct to southern France or even Spain.

Phenomenal flights

Rothiemurchus, 2009

Rothiemurchus, a juvenile male satellite-tagged by Roy Dennis in northern Scotland, set off across the English Channel from Plymouth in a stiff easterly breeze. He drifted west during a long night-time crossing of the Bay of Biscay, and by dawn the next morning, he had missed the north coast of Spain and was 150km (90 miles) out to sea. Had he continued on this course he would have drowned, but he changed direction at first light, making a compensatory change of heading to the south-east. He was now flying against the wind, which would have made the flight across the sea far more arduous, particularly after more than 24 hours already in the air. He was clearly exhausted, because Global Positioning System (GPS) fixes indicated that at 10 o'clock that morning, he had landed on a boat or some other object floating on the sea.

Above: Juvenile Ospreys are susceptible to the effects of crosswinds as they migrate south for the first time. This can lead to them making long ocean crossings.

When he resumed his ocean crossing, Rothiemurchus maintained a south-easterly heading at an altitude of less than 50m (165ft) to reduce the effect of the strong wind. He eventually made landfall at around eight o'clock in the evening just to the south of Porto in Portugal. The sudden change of direction after dawn that morning, and the opportunity to rest on a boat, undoubtedly saved this young Osprey. It is significant that the first signs of this compensatory change of direction occurred only after dawn. The GPS fix at seven o'clock in the morning was 150km (90 miles) west of the Spanish coast, however from the bird's altitude of 47m (154ft) it is possible that he could see the coast from the air, indicating that the sight of land may have triggered this change of heading. In this case, the need to find land took priority over the bird's instincts to continue south-west.

Stan, 2012

Rothiemurchus was lucky. Most young Ospreys that are blown off course and end up out at sea are unable to compensate for their initial mistake. The best example of this is the remarkable migration of Stan, another juvenile male satellite-tagged by Roy Dennis in north-eastern Scotland. Aided by a strong north-easterly tailwind, Stan made rapid progress south, and reached the south-west tip of Portugal only three days after leaving his nest-site in Moray. At this point, an adult Osprey with previous migratory experience would have changed track to travel over the Atlantic to the Moroccan coast. However, Stan, oblivious to what lay ahead and still flying with a stiff north-easterly tailwind, merely continued on the same south-westerly heading he had maintained through Spain and Portugal. He headed out over the Atlantic and eventually made landfall 25 hours later on the coast of Lanzarote in the Canary Islands, after a flight of 1,000km (620 miles) across the sea.

Stan island-hopped across the Canaries the next day, reaching the southern tip of Gran Canaria by evening. At dawn the following morning, the south-easterly wind had strengthened, and so when he set off from his overnight roost, his instinct to migrate south-west was again aided by the wind. By 10 o'clock that night, Stan had already flown more than 700km (435 miles) over the sea, but based on his trajectory he would not reach land until the Cape Verde islands, some 850km (530 miles) to the south-west. Having

Below: Having been blown out into the Atlantic, juvenile Osprey Stan eventually made landfall in Cape Verde, having flown 1,500km (930 miles) in 38 hours of non-stop flight.

already covered 3,500km (2,175 miles) in just six days, this inexperienced Osprey was in serious trouble.

By the time of the next GPS fix at eight o'clock the next morning, Stan had flown a further 460km (285 miles) overnight, but was still around 400km (250 miles) from land. Nevertheless, he showed no signs of letting up, continuing south-west at an altitude of 140m (460ft). Eventually, at around quarter to four on the afternoon of 19 September, the Osprey reached the island of São Nicolau in Cape Verde. He had flown more than 1,500km (930 miles) over the Atlantic in 38 hours of non-stop flight.

Stan remained in Cape Verde for the next week, initially on São Nicolau before island-hopping north-west to São Vicente and then Santo Antão. Sadly, no further transmissions were received, indicating that he may have set off on the same south-westerly trajectory as before and, with the next land some 2,700km (1,680 miles) away in eastern Brazil, drowned at sea. Nevertheless, his remarkable migration to Cape Verde – a flight of more than 5,000km (3,100 miles) in just eight days – demonstrates the incredible capabilities of young Ospreys. Interestingly, Stan departed after a long post-fledging period of eight weeks, suggesting he left in excellent condition with plenty of stored body fat to sustain him during his incredible flight across the sea.

Map showing the migration tracks of three satellite-tagged Ospreys from Scotland.

Adult male **Blue XD** (dark blue line) used a typical autumn route through Europe, crossing the Bay of Biscay and then the western Sahara before reaching his wintering site in the Casamance region of Senegal. The light blue line shows his return migration the next spring.

Rothiemurchus missed the north coast of Spain on his first migration south as a juvenile (orange line), en route to his eventual wintering location in Senegal. He again crossed the Bay of Biscay on his first return flight north (dark red line), but then, the next autumn, avoided repeating the long sea crossing by flying a long dog-leg route down the Atlantic coast of France and north coast of Spain in order to reach a favoured stopover site in Galicia (red line).

Juvenile male **Stan** completed an amazing eight-day migration to Cape Verde from northern Scotland, having been blown off course by easterly winds (green line).

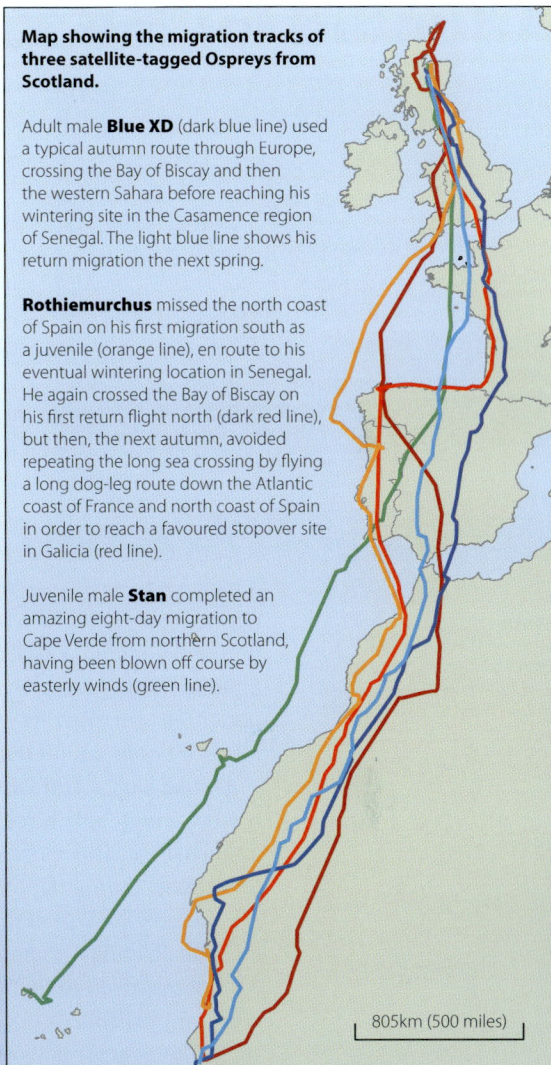

805km (500 miles)

Blue XD migration

Blue XD return

Rothiemurchus first migration

Rothiemurchus return

Rothiemurchus second migration

Stan migration

Adapted for the task

Above: Unlike raptors, which are reliant on thermals to aid their migration, Ospreys migrate by both soaring and flapping flight. This enables them to make long sea crossings.

These long and demanding flights demonstrate the Osprey's extraordinarily ability to undertake long ocean crossings. Most migratory raptors actively avoid long flights over the water because they are dependent on thermal updrafts – which are usually absent over the sea – to aid soaring flight during migration. Flapping flight is hugely costly in terms of energy for heavy birds with broad wings, such as eagles and vultures, and so thermal soaring–gliding is the only way they can migrate to distant wintering grounds. In typical soaring and gliding flight, a bird will circle in a thermal, utilising the rising air to gain altitude. Then, having reached the upper reaches of the updraft, it will open its wings and glide forward, slowly losing altitude until it reaches the next updraft.

Unlike eagles and vultures, Ospreys have long, narrow wings, which reduce drag and make them more aerodynamic. This means that they expend significantly less energy during flapping flight than other raptors, but are still able to exploit thermals when they are available. It is this flexibility in method of flight that means Ospreys are not restricted to flying over land, and can successfully cross vast expanses of ocean.

Adult Ospreys, like juveniles, readily fly across the sea, but they generally attempt crossings only when weather conditions are favourable. Recent satellite-tracking research has shown that adult birds regularly cross the Bay of Biscay from the UK or northern France to Spain during autumn, when north-easterly winds provide valuable tailwind support. In contrast, they are more reluctant to make the return crossing in spring, when headwinds are more common, choosing to fly north along the Atlantic coast of France instead.

Migration strategy

Long, continuous flights over land and sea are often a feature of the migratory journeys of Ospreys, but it is more typical for adults and juveniles to fly 250–300km (155–185 miles) each day. In Europe, where they frequently pass foraging sites, Ospreys tend to fly for one hour per day less than when they are migrating through Africa, sometimes interrupting their day's flight to feed, and usually roosting within a few kilometres of a lake, river or estuary to allow them to fish in the evening or at first light. This tactic, sometimes referred to as fly-and-forage migration, enables some Ospreys to fly directly to their wintering grounds without stopovers.

For example, an adult female Osprey from Rutland Water named 30(05) flew direct to her wintering site on the coast of northern Senegal during five successive autumn migrations, completing the 4,500km (2,796 miles) journey in 11–13 days, always without stopovers. Likewise, Blue XD's 2013 migration to the Casamance region of southern Senegal (see above) – a flight of 5,402km (3,357 miles) – took him 13 days to complete, also without stopovers.

Not all adult birds fly directly to their wintering site, however. Many birds will return to a favoured stopover spot, usually learnt during their first autumn migration, and linger there for a period of a few days to several weeks. Satellite-tracking and ringing studies have shown that individual Ospreys return to the same stopover site on successive journeys. These sites are thought to be an essential part of the navigation process, and become 'goal areas' that the birds aim for on each migration. For example, Nimrod, another of the adult Ospreys

Above: Beatrice, an adult female Osprey satellite-tagged by Roy Dennis in Scotland, had a regular stopover site beside the River Adour in southern France, which she used every spring.

Above: The Sahara presents a demanding challenge for adult and juvenile Ospreys alike, and is one of the main hazards of migration for European birds.

satellite-tagged by Roy Dennis in northern Scotland, returned to Île d'Oléron (just off the French coast near La Rochelle) during three successive autumn migrations, stopping off for between nine and 19 days. Another tagged bird, Beatrice, remained beside the River Adour in southern France for up to three weeks on every spring migration, despite wintering only in Spain. Staying on at a favoured location in this manner enables adult birds to fish and roost in places well known to them, and provides them with an opportunity to increase their fat reserves before they continue their journey.

As at the beginning of autumn migration, adult birds will often make a very long flight after a prolonged stopover. In 2008, Nimrod took advantage of favourable tailwinds upon departure from Île d'Oléron, flying a staggering 2,306km (1,433 miles) in 34 hours to the Western Sahara, including a flight across the Atlantic between Huelva in south-west Spain and Agadir in southern Morocco. It is not clear what prompts some birds to incorporate lengthy stopovers into their journeys and others to fly directly to their wintering grounds, but experiences gained during their early migrations undoubtedly play a crucial role.

Learning from experience

One of the critical findings of satellite-tracking research in recent years has been the understanding of how young Ospreys learn from experience. We know that the first migration is driven by a combination of instinct, weather and places chanced upon during the journey, but subsequent travels are far more predictable. Most first-year Ospreys that are satellite-tagged do not survive long enough to return to their natal site – mortality is as high as 70 per cent during the first two years of life – but those that do survive provide a fascinating insight into how learning shapes migration. In the UK, the best example of this is Rothiemurchus. As described earlier, this male Osprey from Strathspey in northern Scotland, who was satellite-tagged by Roy Dennis as a chick, endured a difficult first migration in which he was blown out into the Atlantic and nearly died off the Portuguese coast. Having made it to land, he remained in Portugal for a month, and then eventually continued south to Senegal, where he stayed for the next 18 months.

When Rothiemurchus returned north for the first time in May 2011, he made excellent progress and reached Europe just seven days after leaving Senegal. As he flew north through Iberia, he veered distinctly to the west, travelling to Galicia in north-west Spain and lingering at the Sor Estuary for just over a week. Then, just as in autumn 2009, he flew directly across the Bay of Biscay, eventually making landfall to the east of Plymouth at Hope Cove, after flying 830km (515 miles) in 28 hours over the ocean. He came ashore at almost precisely the same location that he had departed from in September 2009.

Having made it back to the UK for the first time since his first migration, Rothiemurchus spent the summer exploring Scotland and northern England, visiting his natal nest in June and July, and ranging over a vast area.

Above: Rothiemurchus was satellite-tagged by Roy Dennis as a juvenile and his subsequent travels provided a fascinating insight into how young Ospreys learn from experience during successive migrations.

Eventually, after a summer of exploration, he headed south again in early September. He appeared to have learnt from his previous two migrations because rather than flying across the Bay of Biscay, he followed the French coast instead. He passed Biarritz but then, rather than travelling south through Spain, he continued to follow the coastline, heading due west through the Basque Country, Cantabria and Asturias to the Sor Estuary. He was now back at his spring stopover site, but his dog-leg route around the Bay of Biscay was 450km (280 miles) further than a direct flight across the sea. After a short flight 45km (28 miles) south, Rothiemurchus then spent just over a week beside the River Eume, another place he had lingered at in spring.

In a study of Swedish Osprey migration, Thomas Alerstam and colleagues found evidence for the existence of up to three goal areas, and suggested that adult Ospreys migrate by using different homing processes in succession as they pass through these sites, deviating from the most direct migratory track to check in at each of them. That was certainly the case with Rothiemurchus, whose desire to return to Galicia, but to avoid a direct flight across the Bay of Biscay, added significant distance to his migration.

Above: It usually takes Ospreys a minimum of four days to cross the vast and desolate expanses of the Sahara.

Rothiemurchus's transmitter continued to log data for a further three years. During this period, he continued to visit the Sor Estuary during every migration, sometimes flying across the Bay of Biscay, but on other occasions using the dog-leg route through France and Spain, behaviour that could be traced back to his early migratory experiences. There are numerous other examples of satellite-tagged and colour-ringed Ospreys returning to the same sites each year, confirming the suggestion of Thomas Alerstam and his Swedish colleagues that 'goal areas' play a key role in Osprey migration.

Crossing the Sahara

Covering around 8 million square kilometres (3 million square miles), the vast and desolate expanses of the Sahara present a demanding challenge for adult and juvenile Ospreys alike. Ringing studies have shown that the bulk of European Ospreys winter south of the Sahara, but until the first Ospreys were satellite-tagged in the late 1990s, very little was known about the routes they used to cross the desert. The first documentation of flights across the Sahara was made by Nils Kjellén, Mikael Hake and Thomas Alerstam, a team of Swedish scientists who tracked two adult females from Sweden. One of these birds crossed the desert between Algeria and Burkina Faso, en route to the Ivory Coast, resting at night and flying only during daylight hours.

Since this groundbreaking Swedish study, knowledge of Osprey migration has increased dramatically, most notably with the recent advent of GSM technology. Ospreys migrating south from the UK tend to use a westerly route to traverse the Sahara, setting out across the desert in southern Morocco and then continuing south through Western Sahara and Mauritania, before

Above: Ospreys usually confine their flight over the Sahara to daylight hours, when they can use thermal updrafts to aid their progress across the unforgiving terrain.

arriving in northern Senegal four to five days later. Birds from other populations in western Europe, including France and Germany, use the same route, while those that start their journeys further east – such as Finnish and Estonian Ospreys – usually make a more easterly crossing from Libya or Egypt, south towards either central or eastern Africa. There is variation, however, and some satellite-tagged Finnish birds have been found to use the western route from Morocco to Senegal.

The most comprehensive analysis of the dangers associated with crossings of the Sahara was undertaken by Roine Strandberg and his team at Lund University in Sweden who found that deaths linked to Saharan crossings constituted about half of the annual mortality of juvenile Ospreys. The risk of starvation is high, particularly for juveniles that begin the crossing in poor condition. Ospreys must also negotiate dust storms, which have increased dramatically in the last 50 years. These storms may force migrating Ospreys off their preferred course, and in extreme circumstances may result in aborted crossings or even death.

The risks associated with crossing the Sahara mean that Ospreys must strike a balance between minimising time spent in the desert and conserving energy during the long crossing. The deployment of GSM transmitters has shown that experienced adult Ospreys migrate almost exclusively using a soaring and gliding flight flight over the desert. They circle in the strong thermals to gain altitude, and then glide onwards, slowly losing height until they reach the next updraft. Minimising the need for flapping flight helps them to conserve valuable energy, and so they tend to limit active migration to the middle part of the day, when thermal updrafts are available.

Above: Ospreys migrating through the Sahara are also at risk due to a lack of safe roost sites. It was this that led to the death, on the northern edge of the desert, of Osprey 09(98), from Rutland Water.

Risky roosts

Aside from the dangers of strong winds and dust storms, another risk associated with desert crossings for migrant Ospreys is a lack of safe roost sites. Analysis of satellite imagery indicates that many migrant Ospreys are forced to roost on the desert floor, which significantly increases their chances of predation. This was exemplified by the fate of 09(98), an adult male Osprey that was satellite-tagged at Rutland Water in 2011. Having raised two chicks in 2012, 09(98) set out on migration on 5 September and made rapid progress south through France and Spain, averaging more than 400km (250 miles) per day. By the evening of 10 September, he had reached a ridge on the northern edge of the Sahara, and settled among large rocks and boulders for the night. We hoped that 09(98) had found a relatively safe place to roost, but the next batch of data from the satellite transmitter three days later suggested he was still in the same location, which was a worrying sign.

Farid Lacroix, an ex-search and rescue helicopter pilot living in Agadir, Morocco, offered to help investigate. The next day he drove five hours south from his house. Using his GPS to guide him, Farid trekked up to the ridge in searing desert heat. As was feared, 09(98) had been killed, almost certainly predated by a mammal predator while he roosted on the ridge. It demonstrated that even for an experienced adult Osprey such as 09(98), a 14-year-old that had completed 26 previous migrations, the Sahara presented a genuine danger.

Below: Ex-search and rescue helicopter pilot Farid Lacroix trekked up a remote ridge in the Sahara to find the remains of 09(98), who is likely to have been predated, perhaps by a mammal predator such as a jackal, while roosting.

A winter home

Above: Many European Ospreys spend the winter on the west coast of Africa.

Satellite-tracking studies have confirmed that a small but increasing number of Ospreys from the UK and elsewhere in northern Europe winter in Spain and Portugal. Most, however, continue south to sub-Saharan Africa, before wintering anywhere from southern Mauritania to Ghana. Once there, the adults and juveniles display very different behaviour. Adults head directly to a wintering site that they are faithful to each year, while juveniles must embark on the challenge of finding somewhere to settle.

Wintering juveniles

After crossing the Sahara, young Ospreys using the western route arrive in northern Senegal. The Senegal River and nearby wetlands such as the vast Djoudj National Bird Sanctuary provide the first fishing opportunities for several days. Some juveniles linger here, while others continue south, encountering adult Ospreys at all the best wintering sites along the beaches between

Below: Île des Oiseaux in the Sine-Saloum Delta in Senegal attracts large numbers of wintering Ospreys from Europe.

St Louis and Dakar, and among the 1,800 square kilometres (700 square miles) of shallow water, inter-tidal mangroves and savanna woodland of the Sine-Saloum Delta. If they attempt to settle in these areas, juveniles are often chased off by adult birds that may have been returning to the same site for a decade or more. This prompts some juveniles to continue south towards the Gambia, the Casamance region of southern Senegal or Guinea-Bissau. Others will retreat into less favourable marginal sites – perhaps an ephemeral waterbody, or a backwater of one of the innumerable rivers that flow through the region.

Above: Discarded fishing nets pose a threat to migrant Ospreys and a range of other coastal birds in West Africa.

If a young Osprey is already in poor condition from its journey, moving to areas where fishing is more difficult can be fatal. It must also be wary of crocodiles, jackals and feral dogs, all of which will readily snatch an unsuspecting juvenile Osprey that is floundering in shallow water or perched on the ground. Discarded fishing nets that litter beaches along the West African coast pose a threat, too, and in some locations Ospreys are intentionally killed, with naive young Ospreys especially at risk.

Below: Juvenile Ospreys are often chased away from the best wintering sites by adults, which return to favourite spots every year.

Not all young Ospreys are chased away from the best sites. Some – especially those arriving in good condition – are able to hold their own and linger at places like Djoudj National Bird Sanctuary and the Sine-Saloum Delta for the early part of the winter. However, it is rare for juveniles to remain at the same site for any longer than a few months after first arriving in West Africa, as the urge to move on and explore usually compels them to set off again. This urge helps them to develop knowledge of potential wintering sites, but may also get them into strife. Leaving

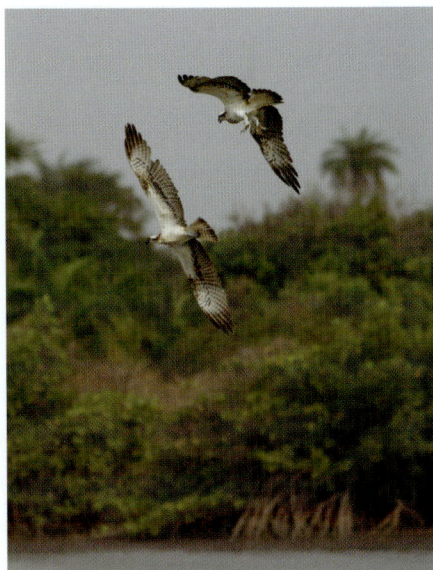

Sudden death

The fate suffered by some juveniles on their wintering grounds is exemplified by the story of a juvenile female known as Fiddich, that was satellite-tagged by Roy Dennis in Scotland in 2012. After arriving in Senegal, Fiddich initially settled beside a river to the south-east of Lac de Guiers. She remained there for much of the early part of the winter, but in mid-December she moved to 100km (62 miles) north-west to Djoudj National Bird Sanctuary.

Fiddich spent two months at Djoudj, before returning briefly to her initial wintering site beside the river near Lac de Guiers. However, it would appear that the river had begun to dry up in the interim because the next day, she flew 300km (185 miles) south-east, before eventually settling beside the Gambia River in eastern Senegal.

After she had been living beside the Gambia River for two weeks, Fiddich's transmitter suddenly started to give signals from a village 2km (1.2 miles) away, suggesting she had been caught in fishing nets or intentionally captured. The transmitter continued to send data from the village, so on 12 April, Gambian bird guide Junkung Jadama, along with Colin Cross from Kartong Bird Observatory, made a special overland trip to try and investigate. Upon arrival in the remote village, Junkung and Colin found the locals to be understandably suspicious about their sudden appearance and questions about the Osprey, and unfortunately they were unable to learn exactly what had happened. Nevertheless, it was clear from the satellite data that Fiddich had died.

a place where they have lived untroubled for several months has pitfalls, and many satellite-tagged birds have died after moving to new sites in late winter or early spring (see box, above). Those that do survive benefit from the adult birds returning north to their breeding grounds in the spring. Most juvenile Ospreys remain in West Africa for the whole of their second year, and the departure of the adults helps them become established at their chosen site. By the time the adults return the following autumn, they are much better prepared to hold their own.

Below: Adult Ospreys fish once or twice per day when at their wintering site.

Wintering adults

Unlike juveniles, adult Ospreys head straight to a known wintering site in the autumn and usually remain there throughout the winter. Satellite tracking has shown that they settle into a predictable daily routine, roosting in the same location each night and favouring the same perching and fishing spots each

day. In fact, some adult birds live in an area as small as 1 square kilometre (0.4 square miles) for six months, which can lead to them being fiercely territorial, particularly towards newly arrived juveniles. However, in areas where food is easy to come by, such as Île des Oiseaux in the Sine-Saloum Delta, adults often perch together during the day.

The aptly named Île des Oiseaux, a long sandy island lying at the mouth of the delta, is surrounded by shallow, fish-rich tidal waters. Two dozen or more Ospreys will often congregate on the sand with Blue-cheeked Bee-eaters (*Merops persicus*) zipping around overhead, and then flop into the water once or twice a day, usually on a falling tide, to catch an easy meal. Life is so straightforward here that there is little aggression between wintering adults, and the Ospreys even allow Slender-billed Gulls (*Chroicocephalus genei*) and Ruddy Turnstones (*Arenaria interpres*) to steal pieces of fish as they eat their catch on the sand. For these birds, life could not be further removed from the experiences of juveniles such as Fiddich (see box, page 66). For young Ospreys, the challenge of establishing a wintering site, particularly after their arduous first migration, is undoubtedly one of the most challenging aspects of the first year of their life. It is only after they have become established at their favoured site that life in winter becomes as simple as it is for adult Ospreys on Île des Oiseaux.

A History of Ospreys in Great Britain

Globally, Ospreys are one of the most widely distributed bird species, and yet compared with historical times, their current range is restricted, including in Great Britain. The Osprey's fishing prowess has been admired for centuries, but the species' penchant for raiding fishponds led to widespread persecution and eventual extinction in England and Wales by the mid-19th century, leaving only a tiny, isolated population in Scotland. Happily, Ospreys are now recovering well, thanks to more than half a century of intensive conservation work that has seen the Scottish population rise to more than 300 pairs, and the species return to both England and Wales.

A drastic decline

There is extensive historical evidence that Ospreys were once a common sight across the whole of the British Isles. The birds were indeed known as highly skilled hunters, but their habit of taking fish from ponds (or stew ponds as they were often called) in the grounds of large country houses and monasteries led to widespread and relentless persecution. The naturalist William Turner wrote in *The Principal of Birds of Aristotle and Pliny* (1544) that 'The Osprey is much better known today to Englishmen than many who keep fish in stews would wish; for within a short time it bears off every fish.' It is clear that at a time when fish was eaten every Friday on religious grounds, the Osprey was seen as a pest, and keepers of stew ponds would have made every effort to eradicate them.

During the 18th and 19th centuries, the killing of birds of prey reached extreme levels, and the Osprey fell into steep decline across much of Britain. The last English pair attempted to breed near Monksilver in Somerset in 1847, but one of the birds was shot by a gamekeeper. With that

Opposite: The Osprey nest at Loch Garten has become synonymous with the recovery of Ospreys in the UK over the last 60 years.

Below: An early painting of Ospreys, taken from *The Birds of Great Britain*, published in 1795.

single shot, the fate of the Osprey in England was sealed.

In Scotland, where there were fewer people and less persecution, the Osprey was just about holding on. At the time of the loss of the last English Ospreys, it is thought that only 40 or 50 pairs were remaining in Scotland, including a nest on the abandoned castle at Loch an Eilein in Strathspey. Many of the remaining nests were in the remote western half of Scotland, away from more populated areas in the east. Nevertheless, as the species became increasingly rare a new threat emerged. There was now a lucrative market for both Osprey eggs and taxidermy specimens. Ironically, travelling naturalists went to great lengths to kill the remaining birds and to take their eggs, giving no thought to the implications of their actions.

By the end of the 19th century, there was just a handful of pairs remaining in Scotland, but, at last, some enlightened individuals recognised the severity of the situation. As Roy Dennis writes in his excellent book *A Life of Ospreys* (2008), the Grants of Rothiemurchus at Loch an Eilein, and the Camerons of Lochiel at Loch Arkaig, made valiant attempts to protect the birds on their land. Sadly, successful breeding was very rare, and the last known pair bred near Loch Loyne in 1916.

Above: Ospreys attempted to breed at Loch Arkaig in 1908 and a lone male returned each year until 1913. The Camerons made valiant attempts to protect the Ospreys on their land.

For many years it was accepted that, as in England, Ospreys had become extinct in Scotland, but after extensive research Roy Dennis now believes that occasional breeding occurred there between 1916 and 1954. Evidence discovered by Roy indicates that Ospreys bred in Aberdeenshire in 1925 and 1926, in Galloway and Loch Luichart between the wars, and at Loch Garten from the early 1930s. In some years, lone birds would have been present at nests, before eventually being joined by a Scandinavian immigrant or a returning Scottish youngster, thereby enabling breeding to continue.

The recovery begins

The efforts of the Grants at Loch an Eilein and Camerons at Loch Arkaig made them the pioneers of the tremendous conservation effort that has gone into helping the Osprey recover in both the UK and around the world, particularly since the mid-20th century. It is now clear that Ospreys did not become extinct in Scotland, but it is only thanks to the efforts of innovative and proactive conservationists in the 1950s and 1960s that the population slowly began to recover.

When ornithologist Desmond Nethersole-Thompson found a pair of Ospreys with two young at the south end of Loch Garten in 1954, the nest was kept a strict secret. In the 1950s, egg-collecting was still a common practice and posed a significant threat to Britain's rarest breeding birds, including Ospreys. Sadly, when the pair returned to the same nest in 1955, the eggs were stolen, and the next spring the nest of a pair of Ospreys on the nearby Rothiemurchus estate was also robbed. It was clear that action was required if the species was to have any chance of making a recovery.

Recognising the grave threat that egg collectors posed, George Waterston, the director of the Royal Society for the Protection of Birds (RSPB) in Scotland at the time, organised a team of wardens to protect the breeding Ospreys. However, in 1958, despite their best efforts, an egg collector still managed to raid the Loch Garten nest one wet night in early June. Unperturbed, the group was reassembled the following spring under the guise of Operation Osprey, and this time the area around Loch Garten was designated as a protected bird sanctuary, making it an offence to approach the nest without permission. The nest was watched day and night, and finally, on 8 June, the behaviour of the birds changed, indicating that the first egg had hatched. Sure enough, a few days later it was possible to make out the heads of three chicks for the first time.

Below: The return of Ospreys to Loch Garten in the 1950s was the start of the recovery of the species in the UK. George Waterston (left) organised this group of volunteers to monitor and preserve the nest.

Conservation visionaries

By this stage, it was proving impossible to keep the Loch Garten nest a secret, so George made what turned out to be a visionary decision: to publicise the good news and to encourage people to come and view the Ospreys from a unique observation point. Over the next two months, 14,000 people took up the invitation, thereby providing the first example of ecotourism in the UK, and a model still copied at many Osprey sites today.

In 1960, aspiring young conservationist Roy Dennis was invited by George to become one of the RSPB's dedicated Osprey wardens at Loch Garten. Roy spent the summer protecting the nest and welcoming visitors to the observation point, where they were able to watch two chicks fledge successfully. It began a life of work with Ospreys for Roy, who without a doubt has gone on to do more to help and conserve the species than anyone else in the UK and across Europe.

Like George, Roy is a visionary, and he has always advocated a proactive approach to nature conservation. In those early days, it was essential that any new nests were protected to deter egg thieves, and all manner of efforts were made, from barbed wire around the base of nest trees to the installation of electronic warning systems. Although some nests continued to be robbed, these early efforts ensured that the Scottish Osprey population began to recover.

Forty-four young Ospreys were reared in Scotland between 1954 and 1970, and in spring 1971 Roy returned to the Highlands to resume his work with the birds after seven years as the warden of Fair Isle Bird Observatory. By now, Ospreys had spread into Moray, where five pairs were breeding, and also to the newly acquired Scottish Wildlife Trust nature reserve at Loch of the Lowes in Perthshire. Like Loch Garten, Loch of the Lowes would later become a famous place for Osprey ecotourism.

Above: The recovery of Ospreys in Scotland and the rest of the UK has been greatly aided by the inspirational work of two conservation visionaries, George Waterston and Roy Dennis (left).

Monitoring nests and ringing chicks

In the 1960s and 1970s, Roy and colleagues established a pattern of nest monitoring that continues to this day, making the UK Osprey population one of the best studied anywhere in the world. The Osprey is protected as a Schedule 1 species under the Wildlife and Countryside Act 1981, and so all monitoring work is undertaken under special licence.

The first visits are made to nests in late March or early April to check for occupancy, and to attempt to identify individual birds from rings or distinctive plumage features. If Ospreys are present, incubation usually begins two or three weeks later; clutch size is sometimes checked using a mirror fixed to the top of an aluminium pole. Subsequent visits in late May and early June enable fieldworkers to establish if chicks have hatched, and to count the number of young. If there are chicks in the nest, then they are ringed – under licence – at around five to six weeks of age. The first Osprey was ringed in Scotland in 1966 by Douglas Weir, and Roy subsequently took responsibility for the ringing programme in 1971. He continues in this role today and, to date, more than 3,000 young Ospreys have been ringed at nests in Scotland and, more recently, England and Wales.

On a ringing day, Osprey chicks (which unlike many raptors are usually very docile) are collected from the nest and lowered to the ground in a bag, while their parents circle overhead. Most adult Ospreys keep a watchful eye

Above: Long poles with a mirror fixed to the top are sometimes used, under licence, to check the contents of Osprey nests.

Above: Young Ospreys are ringed at five to six weeks of age with a metal BTO ring on one leg (left), and a colour ring on the other (right).

Below: Roy Dennis (right) and Ian Perks ringing young Ospreys. Roy has coordinated the UK Osprey colour-ringing scheme since the early 1970s.

on proceedings without showing any aggression, but on rare occasions, one of the breeding pair will dive-bomb the person climbing to the nest. Once on the ground, the chicks are weighed and measured before being fitted with two different leg rings: a metal British Trust for Ornithology (BTO) ring on one leg; and a larger colour ring on the other leg, which enables individual Ospreys to be identified at a distance. In the early years, a single digit on the colour ring was used to determine each bird but, as time progressed and more birds were ringed, it was necessary to use two- and, more recently, three-digit inscriptions. This long-term research, which has enabled the fortunes of thousands of individual Ospreys to be followed, has provided fascinating information on the movements of Ospreys, and greatly enhanced our understanding of how, when and where young Ospreys enter the breeding population.

It is a privilege to observe young Ospreys at such close quarters, and ringing provides an excellent opportunity to share that thrill with landowners and others who have helped protect their local birds. Over the years, the support of local people – including landowners, foresters, gamekeepers and farmers – has been key to the recovery of Ospreys in Scotland and elsewhere in the UK.

The recovery continues

The Scottish population continued to increase both numerically and geographically through the 1970s, and by the end of the decade, there were breeding pairs in Aberdeenshire, Easter Ross and Angus, in addition to those in Strathspey, Moray and Perthshire. In 1980, the number of young reared reached 250. The careful monitoring, protection of nests and liaison with local people by Roy Dennis and others played a crucial role during this period, but even so, some nests continued to be robbed, and in other years poor weather resulted in nest failures. Nevertheless, there was now a clear upward trajectory that would continue through the 1980s and into the early 1990s.

The population reached a significant milestone in 1996, when for the first time in several centuries there were more than 100 breeding pairs in Scotland. The geographical expansion, however, was proving to be relatively slow, with most new pairs becoming established in areas already holding Ospreys. Even so, 1996 did see the first pair breed in the Scottish Borders.

Below: The Scottish Osprey population reached 100 pairs in 1996.

Staying close to home

The 30th anniversary of the first Osprey being ringed in Scotland was celebrated in 1996, and it was thanks to colour-ringing that it was possible to understand why the geographical expansion of the species was slow. By comparing where an Osprey first bred with the location of its natal nest (the site where it hatched), it was possible to calculate the average natal dispersal of both males and females within the population. This provided some fascinating information. The data showed that males tend to be highly philopatric, usually preferring to nest within a few kilometres of their natal site. Females tend to be less site faithful, with some individuals moving 100km (60 miles) or more from their natal site. However, because it is the males that build nests, their reluctance to establish nests in new areas limits the geographical expansion of the population.

Over the years, this pattern of dispersal meant loose colonies of Ospreys formed in the strongholds of Strathspey, Moray and Perthshire, but the birds remained absent from other areas with suitable habitat, even though migrant Ospreys passed through every spring and autumn. Occasionally, a pioneering bird would confound

Below: A male Osprey delivering a fish to his mate on the nest. Research has shown that male Ospreys usually return to breed close to their natal site, whereas females often disperse much further.

the usual pattern and build a nest in a new area, and, with luck, attract a mate. It was clear, however, that these outlying colonies had a chance of becoming established only if these pioneering individuals were soon joined by other Ospreys. The annual survival of adult Ospreys in the Scottish population is around 90 per cent. However, if one of the birds in a new pair fails to return in the spring, and there are no other birds to take its place, the lone individual will be forced to move back to one of the established areas, and the colonisation attempt will fail.

Despite the difficulties of colonising new areas, the Scottish Osprey population continued to increase during the early part of the 21st century. A total of 153 pairs reared more than 200 chicks for the first time in 2001, and by 2005 breeding pairs were established in Lochaber in the western Highlands and Galloway in south-west Scotland. Today, there are close to 300 pairs widely distributed across most areas, and, as we will learn in the next chapter, Ospreys have now returned to England and Wales too, thanks to a proactive conservation effort. This is a tremendous success story and testament to the hard work, dedication and vision of George Waterston, Roy Dennis and others who have dedicated their lives to monitoring and protecting this majestic bird.

Osprey Conservation

Over the past 50 years, a great deal of conservation effort has gone into helping Osprey populations recover from the effects of persecution, habitat loss and harmful insecticides. It is clear that the Osprey is a species that responds well to proactive support, as illustrated by the success of artificial nest-building and reintroduction projects. Osprey translocations were first undertaken in the United States, and they have since been replicated in several parts of Europe, helping to restore Ospreys to areas where they had been lost for decades or even centuries.

Building artificial nests

In the early years of the Osprey recovery in Scotland, the protection of nests against egg collectors was of utmost importance. However, the introduction of custodial sentences for those convicted of the crime led to a significant reduction in the number of nests of Ospreys and other rare breeding birds that were robbed. Thankfully, egg-collecting is now far less of a threat than it was in the 1950s and 1960s, although a few isolated cases do still occur.

With the threat of egg-collecting now significantly reduced, the emphasis of Osprey conservation work has shifted. During the middle to latter part of the 20th century, it became clear that the Osprey is a species that responds well to decisive, proactive intervention.

One of the very best ways of helping Ospreys is to build artificial nests. Roy Dennis constructed his first Osprey nest in Scotland in 1960 using an old cartwheel. Although that nest was never used by Ospreys Roy, and colleagues have since become adept at building nests, and it has proved an essential means by which to maximise the breeding success of Ospreys in existing colonies, and to help them spread to new areas.

Opposite: Ospreys build their nests in the top of prominent trees, but they will also readily use artificial nests. This has been key to their recovery in the UK and around Europe.

Below: Ian Perks climbing to an artificial nest in a tree in Scotland.

Above and right: Artificial nests can be built on telegraph poles if there are no suitable trees. Electrical companies often kindly assist with the installation of these nests.

Above: Artificial nests are lined with soil, turf and moss to create a flat platform, almost level with the top of the sticks. This replicates how an Osprey nest looks at the end of the breeding season.

There are two keys ways in which building artificial nests can help Ospreys. First, young, inexperienced Ospreys are prone to building poorly constructed nests in precarious positions. There is often a danger that these nests will collapse during the breeding season, leading to the loss of their eggs or chicks. Roy and colleagues in Scotland became skilled at identifying nests at risk of collapse, and replacing them with secure nests during the winter. They also repaired nests that had been damaged by storms. Even experienced adult Ospreys may waste valuable time and energy repairing nests when they return in the spring, and if the damage to the tree is particularly severe, they may be forced to move to an entirely new location. In established Osprey colonies it is vital to check all nest-sites during late winter to ensure that any remedial work can be carried out in good time before the birds return in late March. This is the work that Roy pioneered in Scotland, and that is now carried out in many areas with breeding Ospreys.

Artificial nests have also been an excellent way to encourage Ospreys to spread to new areas. In recent

Above: Artificial nests on telegraph poles have been used with great success at Rutland Water.

years, satellite-tracking research and sightings of colour-ringed Ospreys have shown that when two-year-old Ospreys return to the UK for the first time, they wander over vast areas. These exploratory flights are not random; instead, they help subadult Ospreys to map the location of established nests. Careful analysis of the satellite-tracking data has shown that subadult birds visit Osprey nests throughout Scotland, while high-definition webcams often capture intrusions by non-breeding colour-ringed Ospreys. Research has shown that first-time breeders prefer to take over established nests rather than build their own from scratch, and so, by checking out the locations of successful nests, subadult Ospreys are able to identify potential breeding sites.

This knowledge is particularly crucial for female Ospreys, because they have a tendency to nest further away from their natal site than males. As discussed on page 76, male Ospreys usually return to breed close to their natal site, and this limits the geographical expansion of Ospreys. However, this problem is further exacerbated by the fact that young Ospreys prefer to nest close to other Ospreys, and to take over established nests rather than build their own. In many parts of the UK where there is suitable breeding habitat, there is no recent history of breeding Ospreys and therefore no old nests to settle on. In such a scenario, the erection of artificial nests enhances the chances of Ospreys becoming established in new areas. Over the years, the construction of artificial nests in Scotland has been a critical factor in facilitating the species' expansion from the strongholds in Strathspey and Moray. Furthermore, it also aided the recolonisation of Cumbria in 2001 and Kielder Forest in Northumberland in 2009.

Below: Artificial nests encouraged Ospreys to spread south from the Scottish Borders to Kielder Forest in 2009.

The ideal nest

So, what makes an excellent artificial Osprey nest? The first requirement is that it looks as much like an active Osprey nest as possible. That might seem an obvious point to make, but tying a few sticks to a platform is not enough. When building an artificial nest, it is necessary to think about what an Osprey nest looks like at the end of the breeding season. In the early part of the year, there is a distinct nest cup that helps to protect the eggs and small chicks, but as the days and weeks progress and the chicks near fledging, the adults bring in clumps of nest lining in order to create a flat platform from which the youngsters can launch themselves into the air. This means that by the time the family departs in early September, the nest lining

Below: Roy Dennis has pioneered artificial Osprey nest-building in the UK.

is almost level with the top of the sticks. The male's first job in the spring is always to scrape out the nest cup, so when building an artificial nest, it is crucial to line it with a combination of turf, moss and soil. This will ensure that it resembles an established nest to the Osprey when it first returns in spring.

It is also necessary to build a large structure, around 1m (3ft) in diameter. As discussed earlier, Osprey nests can be used for many generations, and so the general rule when building artificial nests is: the bigger it is, the better. It is essential to give careful consideration to the location as well. Ospreys prefer a clear, uninterrupted view and so it is important that the nest is built in a prominent location. This could be on the top of a tree or on a telegraph pole. In some situations, such as on saltmarshes or other areas with no trees, a very low telegraph pole will suffice, but in other places, with tall trees, it is essential to build the nest as high as possible. It does not need to be on the water's edge – many Osprey nests are several kilometres from the nearest water – but it does need to be within a relatively short flight of suitable fishing grounds.

Above: Reserve officer, Lloyd Park, putting the finishing touches to an artificial nest at Rutland Water. The team at Rutland have become experts at building nests.

Translocations

Building artificial nests has been a successful way of encouraging Ospreys to spread through Scotland and into northern England, but greater geographical jumps require something altogether more proactive. The Scottish population reached 100 breeding pairs in 1996, but the majority were still confined to the strongholds of Strathspey, Moray and Perthshire. Migrant Ospreys were now being seen at reservoirs and gravel pits in southern England on a regular basis, and some forward-thinking conservationists had already built artificial nests in the hope of enticing passing Scottish birds to stay and breed. The big problem, however, was that there were no Ospreys that regarded England as home.

In early 1995, Tim Appleton, the reserve manager at the Leicestershire and Rutland Wildlife Trust's Rutland Water Nature Reserve, asked Roy Dennis to come and build five artificial Osprey nests at the large reservoir in central England. Roy did so on a wet and windy day in mid-March, and during the day discussed the idea of translocating young Ospreys to the site rather than waiting for the nests to be colonised naturally, which – given the experiences in Scotland – may have taken many decades.

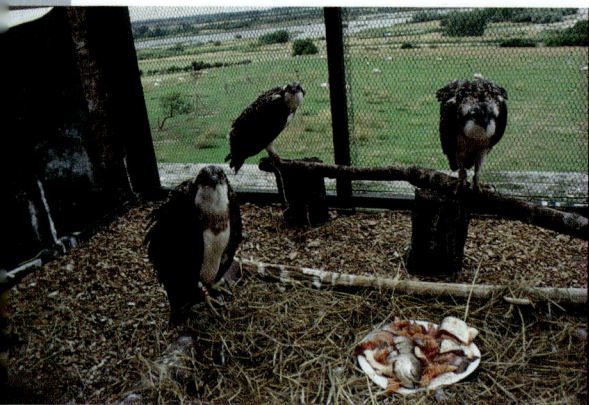

Above: The translocated Ospreys were kept in special pens – or hacking cages – at Rutland Water for two to three weeks before release.

Tim, like Roy, is an advocate of a proactive approach to conservation, and he leapt at the idea, with the full support of Anglian Water, the reservoir's owners. Nothing like it had been attempted in Europe before, but there was a precedent for the work in North America, where Osprey populations were decimated during the mid-20th century by the widespread use of harmful insecticides such as dichlorodiphenyltrichloroethane (DDT). These insecticides entered the food chain and led to the thinning of eggshells and the subsequent failure of nests. Populations of Ospreys across the United States were severely affected, and it was only once these insecticides were banned that Osprey populations began to recover, aided in some cases by translocation projects.

Translocation – or hacking, as it is often called – involves moving chicks from well-established colonies and releasing them in areas where previous Osprey populations have been lost. Juvenile Ospreys imprint on an area when they are first flying, and so moving chicks at six weeks of age, just before they fledge, ensures that once they are old enough to return – usually at two years of age – they head back to the translocation site, rather than the place where they originally hatched. Thus, by using the Osprey's own tendency to return to its natal site, it is possible to restore the species to parts of its former range far more quickly than through natural recolonisation. In the United States, the first translocation project was undertaken in the Tennessee Valley in 1979, and subsequent projects began in north-eastern Pennsylvania and New York the following year, and in central Minnesota in 1984. Later projects were also undertaken in New Jersey, Kansas, West Virginia and North Carolina, and it was soon apparent that the species responded very well to this management technique.

Having helped pioneer the translocation of both Red Kites (*Milvus milvus*) and White-tailed Eagles in the UK, Roy was adamant that it was the best means of restoring Ospreys to their former range in the UK and mainland Europe. And so, after lengthy consultations, and having demonstrated that the project fulfilled the various criteria set out by the International Union for the Conservation of Nature (IUCN) for reintroduction projects, a licence was granted by Scottish Natural Heritage to translocate a small number of Ospreys from nests in northern Scotland to Rutland Water. This was a real milestone for the conservation of Ospreys in Europe.

Above: On release day at Rutland Water the front of each pen was lowered to allow the birds to fly out when they were ready.

Rutland Water

Over six summers between 1996 and 2001, a total of 64 juvenile Ospreys were collected by Roy Dennis and colleagues from nests in Highland and Moray, and driven by road to Rutland Water. The initial licence allowed Roy to collect the smallest bird from a brood of two or three, but it was soon apparent that it was far better to translocate larger, more robust individuals. Sadly, four of the eight birds that were translocated during the first year in 1996 died either before or just after release. The revised techniques were much more successful, however, and of the 56 birds translocated between 1997 and 2001, 55 set off on migration. A further 11 birds were translocated and released in 2005, with 10 departing on migration a few weeks later.

Upon arrival at Rutland Water, the birds were housed in specially designed release pens, which after the first year were sited on Lax Hill, giving the birds a commanding view over the nature reserve and the reservoir. They were held in groups of three to replicate conditions in a natural nest, and fed locally sourced trout three times a day. Human contact was kept to an absolute minimum in order to avoid any chance of imprinting, but the birds were monitored from a distance using telescopes and closed-circuit television (CCTV) images. This enabled the project team to monitor each individual's development and decide when they were ready for release, which was usually after two or three weeks in the cages. Before release, the birds were fitted with small tail-mounted radio transmitters to help the team track their local movements, and a few of them were also equipped with satellite transmitters.

Release day was always exciting. The front of each pen was lowered at dawn, to allow the juveniles to fly when they were ready. Some took their chance of freedom within minutes, while others were altogether more hesitant – sometimes to the extent that the front of the pen had to be closed for the night and then reopened the next day. First flights were equally variable. Some birds made short, simple trips to a nearby perch, while others flew much further. One bird even landed on the spire of the local church.

Below: A total of 64 five-to six-week-old Ospreys were translocated to Rutland Water between 1996 and 2001, with a further 11 birds released in 2005. The project was the first of its kind in Europe.

Settling in

At natural nests, adult Ospreys feed their offspring until they set off on migration, and the post-release feeding regime at Rutland Water replicated that with food placed on top of the pens, and on nearby artificial nests two or three times per day. This meant that the juveniles were in the best possible condition when they departed on migration. During the post-release period, the juveniles soon grew in confidence on the wing, venturing further from the release pens on a daily basis. Within two weeks of making their first flight, they were ranging several kilometres each day before returning to the release pens to feed.

During the Rutland translocation, many people were surprised that the young Ospreys didn't need their parents to show them how to catch fish. As on the first migration, the urge to hunt is entirely instinctive for juvenile Ospreys, and it was not long before the translocated birds were making practice dives, with varying degrees of grace, into the water. As happens at natural nests, it was uncommon for these dives to be successful, but they are an essential element of the post-fledging period, helping to prepare the

Below: 08(97) was the first of the translocated Ospreys to return to Rutland Water, in May 1999.

juveniles for the first migration, during which it becomes imperative that they catch fish for themselves.

In late August and early September, as the days grew shorter and night-time temperatures dropped, one by one the juveniles began to set off on migration. Fourteen of the birds were fitted with satellite transmitters, allowing the project team to follow their progress through Europe and then on to West Africa, but for the remaining birds, it was a case of keeping fingers crossed and hoping that they may reappear two or three years later. The satellite tracking demonstrated just how dangerous the first migration could be. Two birds drowned at sea, and some were lost in the Atlas Mountains and the Sahara, while others perished once they arrived at the wintering grounds. It wasn't all doom and gloom though. The tracking demonstrated the remarkable migratory abilities of juvenile Ospreys that had been on the wing for just a few weeks before departure. One of the birds migrated to the Senegal River in a little over three weeks, while another averaged 231km (144 miles) per day in a 22-day migration to northern Senegal. This gave the project team hope that at least some of the translocated birds would make it back to Rutland.

Above: 03(97) returned to Rutland for the first time in 1999 and went on to raise a total of 32 chicks at his nest close to the reservoir. Here he can be seen watching his family from a favourite perch next to the nest.

Many happy returns

Above: The first Osprey chick in central England for more than 150 years fledged from the Site B nest at Rutland Water in 2001.

The first signs of success came in May 1999, when a translocated bird, 08(97) – 08 being the number inscribed in the bird's colour ring and 1997 being the year of release – arrived back at Rutland Water for the first time. Two weeks, later a second bird, 03(97), was seen nearby, and it was clear that the project was working. These male Ospreys regarded Rutland Water, not Scotland, as their home. The next summer, two more translocated males made it back to Rutland Water, and 03(97) built a superb nest on the top of an oak tree on private land close to the reservoir.

In spring 2001, 03(97) returned to Rutland on 28 March, some 19 days earlier than the previous year. It was clear he was now intent on breeding. On 3 April he was joined by a female. She was not ringed, but likely a bird from Scotland. A deformity in one of her eyes suggested she may even have been an old bird that had been ousted from the breeding population by a younger rival female. After a courtship period of several weeks, she laid eggs and then, five-and-a-half weeks later, a single chick hatched. This was a momentous moment for Osprey conservation in the UK and, as were the early successes at Loch Garten (see page 71), it was marked with much excitement. The Osprey was back in central England after an absence of 150 years. That summer, a pair of Ospreys also bred successfully in the Lake District for the first time, aided by the erection of artificial nests, and in Scotland 153 pairs reared a total of 200 chicks. It was clear that, after the drastic declines of the 19th and early 20th centuries, things were looking up for Ospreys in the UK.

Since the success of 2001, the Osprey population in Rutland has become well established. In the early years, the growth of the colony was slower than expected because fewer Scottish females stopped off to breed than was hoped, but the number of breeding pairs has grown gradually over time. In 2007, a wild-fledged Rutland female, 5N(04), raised chicks for the first time, and then in 2013 one of her offspring from 2010, male 11(10), raised three chicks with female 25(10), the first pair of second-generation Rutland Ospreys to breed successfully. It was initially predicted that the reservoir and surrounding area could support eight to 10 pairs of breeding Ospreys, and eight pairs bred for the first time in 2015, the same year that the 100th chick to fledge from the Rutland Water colony made its first flight. And what of 03(97)? He continued to return to his nest-site atop the oak tree for 15 summers, raising a total of 32 chicks. In 2016, he failed to return to Rutland Water, but he had left a legacy that will live on for generations.

Below: 5N(04) was the first wild-fledged Rutland chick to breed successfully, in 2007. She bred with 08(97) in Manton Bay on the nature reserve and raised two chicks.

A Welsh surprise

The success of the Rutland translocation was not restricted to central England. In 2004, two pairs of Ospreys were found breeding in Wales and, remarkably, the males at both sites were identified as birds that had been translocated to Rutland Water, 07(97) and 11(98). Neither had been seen back at Rutland Water since their first migration, but it was significant that they were breeding at almost the same latitude. Sadly, 11(98)'s nest in the Glaslyn Valley in Snowdonia collapsed during heavy rain and high winds, killing the two chicks. However, there was happier news in mid-Wales, where 07(97) raised a single chick with a female that had fledged from a nest on the Black Isle, north of Inverness in Scotland. Their chick was the first to fledge from a nest in Wales for several centuries, and it was becoming clear that the Rutland project had far-reaching and unexpected effects. It is highly unlikely that two Scottish males would have settled to breed in Wales without first being translocated to Rutland Water.

Below: Ospreys first bred in the Glaslyn Valley in North Wales in 2004. The male was a bird who had been translocated to Rutland Water in 1998.

Unfortunately, the mid-Wales pair failed to appear in 2005, but 11(98) and his unringed mate returned to their nest and raised two healthy chicks. Since then, the Welsh population has slowly increased to a minimum of four breeding pairs, and by 2018 at least 85 chicks had fledged. Rutland birds have continued to play a vital role in the establishment of the population. In 2011, a female Osprey from Rutland Water, 03(08), paired with an unringed male on an artificial nest erected by the Montgomeryshire Wildlife Trust at their Cors Dyfi reserve, and went on to raise two chicks. When she failed to return in 2013, her place was taken by another Rutland-fledged female, 12(10), or Glesni, which reared 12 chicks over five summers with the resident male, Monty. Glesni did not return in 2018, and Monty subsequently paired up with a third Rutland female, 3J(13). With other Rutland birds regularly seen in Wales, it is clear that the translocation has completely changed the distribution of Ospreys in the UK. This is further exemplified by the fact that three male chicks from the Glaslyn nest have gone on to breed in northern England, in the expanding populations in Kielder Forest and Cumbria, where the provision of artificial nests has facilitated the natural spread of Ospreys across the Scottish border.

Above: 3J(13), or Telyn as she was later named, is the third female from Rutland Water to breed at the Cors Dyfi nest in mid-Wales.

A move south

Breeding Ospreys were once widespread along the south coast of England, where the old English name for the species is Mullet Hawk. Today, Ospreys are regular visitors in spring and autumn, but despite the provision of artificial nests, there had been no breeding attempts by 2017. With the population at Rutland Water well established, a licence was granted for a second English translocation at Poole Harbour in Dorset. This large natural harbour, with abundant populations of mullet, European Sea Bass (*Dicentrarchus labrax*) and other estuarine fish, is ideally located to act as the nucleus of a new south coast colony, which will link expanding populations in central England, Wales and central France. The first eight juveniles were translocated from northern Scotland during July 2017 in a partnership between the Roy Dennis Foundation, Birds of Poole Harbour and Wildlife Windows. Encouragingly, two of these birds were seen – just 21km (13 miles) apart – at the Sine-Saloum Delta in Senegal in January 2018.

Below: Poole Harbour in Dorset is the site of the second English Osprey translocation project.

A young female from Rutland Water, CJ7, was observed interacting with the translocated juveniles shortly after their release, and returned again in 2018. It is hoped that the translocated birds will encourage other passing Ospreys to linger at Poole Harbour in future years, as observed at other translocation sites around Europe.

Conservation on the Continent

The Rutland Water translocation was the first of its kind in Europe and, like the projects in North America before it, demonstrated that the Osprey responds well to this proactive form of conservation management. The translocation and release techniques were refined over the years at Rutland Water and implemented with great success. This created a model that could be easily replicated elsewhere in Europe.

As in the UK, the current distribution of Ospreys around Europe, particularly in southern parts, is extremely restricted compared with historical times, with persecution, habitat loss and the use of harmful insecticides leading to the loss of breeding Ospreys in many areas. The work at Rutland Water demonstrated that translocation could play a key role in restoring these lost populations.

Above: Translocated Ospreys at Urdaibai in northern Spain. Ospreys have now been translocated to five different countries in Europe, following the success of the Rutland project.

Spain
ANDALUCÍA

In 2003, an ambitious project to re-establish Ospreys in mainland Spain, after an absence there of more than two decades, began in Andalucía. A total of 164 birds were translocated from Germany, Scotland and Finland over the course of 10 summers and released at two sites some 150km (90 miles) apart: Barbate Reservoir in the south, and at the Odiel Marshes, close to the city of Huelva.

The project was an immediate success. In spring 2005, a pair of unringed Ospreys began incubating eggs at a nest 30km (18 miles) away from the Barbate release site. Although not translocated birds, it seemed that they had been attracted by the presence of the released juveniles the previous summer. They laid a single egg, but after 60 days of continuous incubation, it became clear that it was addled and would not hatch. To ensure that the breeding attempt was successful, and at the same time encourage site fidelity in the birds, the decision was made to translocate two German chicks earlier than planned, so that they could be fostered by the Ospreys.

While the necessary arrangements were made – and to ensure that the breeding birds did not lose interest in the nest – a local Black Kite (*Milvus migrans*) chick was placed in the nest. The Ospreys immediately accepted it, and the male soon returned with a fish, and then watched his mate feed their newly adopted chick. They continued to care for the Black Kite until two 10-day-old Osprey nestlings arrived from Germany. At this point, the kite was returned to its nest and replaced by the two Osprey chicks. As before, they were immediately accepted by the adult Ospreys, and both went on to fledge successfully. This cross-fostering experiment was another superb example of how a proactive approach can be of great benefit to Osprey conservation.

Above: A total of 164 Ospreys from Scotland, Germany and Finland were translocated to Andalucia in the south of Spain over the course of ten summers. There are now over 25 breeding pairs.

The first pair of translocated Ospreys – a German male and Scottish female – bred at the Odiel Marshes in 2009, and since then the Andalucía population has gone from strength to strength. It now numbers more than 25 breeding pairs, and the long-term future of Ospreys in southern Spain is secure.

BASQUE COUNTRY

Even though migrant Ospreys pass through northern
Spain in large numbers in spring and autumn, there have
been only isolated breeding attempts in the north of Spain
during the past century. Ospreys bred in Asturias until
the 1960s, and the only other nesting attempts occurred
in the Basque Country in 1973, and at two reservoirs
in Aragon in the late 1990s and early 2000s. Bearing
this in mind, and considering that the nearest breeding
population is located in Orléans Forest in central France,
some 600km (370 miles) north, translocation was deemed
the best means of restoring a self-sustaining breeding
population.

The estuary and marshes of the Urdaibai Biosphere
Reserve on the Basque coast regularly attract migrant
Ospreys, and the site was considered highly suitable for a
translocation project. A total of 60 Scottish Ospreys were
collected from nests in the Highland and Moray regions
between 2013 and 2017, and translocated to Urdaibai.
The first translocated birds returned in 2015, and by 2017
seven males had returned to the area. That summer, one
of the 2013 males, P1, was seen with an unringed female

Above: Sixty Scottish Ospreys
were translocated to the Urdaibai
Biosphere Reserve in the Basque
Country in northern Spain between
2013 and 2017.

on an artificial nest on the river Cubas near Santander, just two days after it had been erected. The pair stayed together for the rest of the summer and then returned to breed for the first time in 2018. Meanwhile, another male bird, N4, also translocated in 2013, set up territory on an artificial nest at Courant d'Huchet Nature Reserve in the Les Landes region of southern France. He attracted a female from Corsica in May and, like the Santander birds, they remained together until September and then returned to breed in 2018, raising two chicks. This demonstrated how establishing Ospreys in the region has the potential to link existing populations of Ospreys in central France, Corsica and Andalucía.

Portugal

A combination of persecution and disturbance resulted in the loss of breeding Ospreys in Portugal in 1997, with the final male bird disappearing in 2002. An initial plan to restore a breeding population through translocation did not receive the necessary support, but a licence was finally granted in 2011 following the successes of the Spanish project in Andalucía. A total of 56 Finnish and Swedish chicks were subsequently released at the vast Alqueva reservoir in eastern Portugal.

Below: Ospreys have returned to breed in Portugal following a translocation of Finnish and Swedish Ospreys to the Aqueva reservoir.

The first successful breeding occurred in 2015, when a pair reared two chicks on the same sea cliffs in Costa Vicentina that were used by the last breeding pair in 1996. Neither bird was ringed, but, as in Spain, it is thought that the presence of translocated birds in the region – at Alqueva and also in nearby Andalucía – helped attract them. That summer, a pair also reared a single chick at the Alqueva reservoir, but it was not possible to identify the adult birds. However, next spring a translocated male from Finland paired with a female that had been translocated to Andalucía from Germany, and reared two chicks. This was significant in that it was the first confirmed breeding by a bird translocated to Portugal, but also because it demonstrated how birds are likely to move between the newly established populations in Andalucía and southern Portugal in future years.

Italy

While most northern populations of Ospreys are migratory, many of the birds that breed around the Mediterranean are either sedentary or make only limited post-breeding movements. By the early part of the 21st century there had been no breeding Ospreys in mainland Italy for more than 50 years, but in centuries past they would have been a familiar sight along the Italian, Corsican and Sardinian coasts, often breeding on the spectacular sea cliffs that are a feature of this part of the Mediterranean. In 2006, a project was started with the aim of restoring a population of Ospreys at Maremma Regional Park on the Tuscany coast. A total of 32 chicks were translocated from nearby Corsica, where 25–30 pairs breed each year, mainly on cliffs in the north-west of the island. The first pair bred successfully in 2011, and a small population is now established at Maremma and on adjacent marshes. In fact, in 2017 two pairs of Ospreys each reared chicks on artificial nests just 1km (0.6 miles) apart at Diaccia Botrona Nature Reserve.

Above: In 2011 a pair of Ospreys raised two chicks at Maremma Regional Park – the first successful Osprey nest in mainland Italy for over 50 years.

Switzerland

Ospreys last bred in Switzerland along the River Rhine in 1914, with the last observation being of a territorial male in 1919. Since then, sightings have generally been confined to migration periods, even though extensive areas of suitable breeding habitat exist in many parts of the country. With natural recolonisation considered unlikely, a five-year translocation project was initiated in the canton of Fribourg, in western Switzerland, in 2015. Six Scottish juveniles were translocated that summer, and in subsequent years birds were moved from Germany and Norway. Excitingly, one of 12 birds released in 2016 was photographed in northern Senegal by John Wright in December that year, proving it had survived its first migration south. The first translocated bird to return, a Norwegian male released in 2016 named Fusée, was subsequently seen back in Switzerland in spring 2018.

Below: The first Ospreys were translocated to Switzerland in 2015.

France

In 2018, authorities granted a licence to translocate Ospreys from central France to the Atlantic coast north of Biarritz. The first 10 birds were moved later that summer. Once established, this new population will help to link existing populations in Orléans Forest and other parts of central France with those in northern Spain and also the Mediterranean region.

The future of Osprey conservation

The translocation projects around Europe have had a hugely positive impact on the distribution of Ospreys, but much work remains to be done. There are still large parts of southern Europe in particular that lack breeding Ospreys, and it is now clear that translocation is the most effective means of restoring the species to these areas. The knowledge and capability to run these projects successfully, and cost-effectively, is now readily available and, just as in North America, further translocations will help the species to recover to historic levels. In 2016, the Council of Europe published a report written by Roy Dennis that recommended this ongoing positive approach to Osprey conservation in Europe.

The conservation of charismatic flagship species such as the Osprey can play a crucial role in helping to raise awareness of the conservation of wetland ecosystems, and to generate broader public support for the protection movement. Ospreys create great interest wherever they occur, so aside from the moral responsibility we have to help a species that was lost entirely through the influence of humans, the return of the Osprey can have many other far-reaching benefits.

Ospreys in Culture

The Osprey is a bird that has been admired for centuries, and as a result, it is referred to in many historical texts, including by Aristotle in the fourth century BC. Today, the species has become something of an icon of conservation and the ecotourism movement, helping to inspire people to take more of an interest in nature, and it is indicative of what can be achieved with a proactive approach. The Osprey has become embedded in our modern-day culture, with aircraft and even sports teams named after this fantastic bird.

As explained earlier, the scientific name for Osprey, *Pandion haliaetus*, is derived from the mythical Greek king of Athens, Pandion I, and the ancient Greek *haliaietos*, *hals* meaning 'salt' or 'the sea' and *aetos* meaning 'eagle'. This was presumably a reference to the Osprey as a bird of coastal habitats, but it is the species' fishing prowess – both in marine and inland waters – that makes it so revered around the world. The words of a Wolof song, sung by Senegalese fishermen, capture the very essence of what makes the Osprey such an icon:

Opposite: The Osprey has been well known for its fishing prowess for centuries.

Below: Fishermen in Senegal are used to seeing Ospreys diving for fish near their boats.

Osprey, the special one,
   fisherman of the sea,
He does not have nets,
   he does not beg for fish,
And he only eats fat fish,
The fisherman and his boat,
The Osprey and his skills,
There will be no lack of fish.

of *Coriolanus*, written by William Shakespeare between 1605 and 1608. In Act IV, Scene vii, Aufidius uses the Osprey as a metaphor to describe Coriolanus's military skills and hold over Rome: 'He'll be to Rome as is the Osprey to the fish, who takes it by sovereignty of nature.' This wonderful line captures the Osprey's fishing prowess perfectly, and also indicates that the birds were a familiar part of the English landscape in the early 17th century, even though they would already have been affected by persecution.

Many writers were fascinated by the Osprey's feet, and another commonly held misconception in medieval times was that the birds had one webbed foot for swimming and one foot with sharp talons to grasp hold of fish. For example, the clergyman William Harrison, in his 1586 *A Description of England*, explains that: 'The Osprey hath one foot like a hawk's to catch hold withal and another resembling a goose, wherewith to swim.' It was not until the late 18th century that this error was identified by John Walcott in Volume 1 of his 1789 *Synopsis of British Birds*: 'Old authors, and Linnaeus after them, erroneously describe the left foot of this bird as semi-palmated; the right, say they, being designed to hold the prey; the left to swim with: the same is related of the Sea Eagle, a bird which, from similarity of manners, has often been confounded with it.'

Walcott's descriptions, as well as those of William Lewin in *The Birds of Great Britain* (1795), were far more accurate than those of medieval writers, but both authors incorrectly suggest that Ospreys would also take waterfowl in addition to fish. Lewin writes, 'This bird frequents the sea-shores, and large rivers, in various parts of Great Britain, and feeds chiefly on fish; plunging with great violence into the water after them; they will also feed on water fowl.' Both Walcott

Below: The Osprey's amazing feet, highly adapted to keeping hold of slippery fish, have fascinated people through the ages.

Left: The accuracy of Osprey artwork improved considerably over the centuries, as this hand-coloured lithograph of an Osprey, taken from John Gould's *The Birds of Great Britain Vol.1* (1873), demonstrates.

and Lewin state that the Osprey 'makes its nest on the ground, among reeds, and lays four or five white eggs'. Could it be that, in this instance, they were mistaking Ospreys for Marsh Harriers (*Circus aeruginosus*)? Without the luxury of today's superb high-powered optics, it would be an easy mistake to make, particularly if the two species were breeding in the same areas, which we know they often did.

Lewin's book was also notable for the fact that it included a very early coloured illustration of an Osprey, finished by hand. Up until this point, all other illustrations had been in black and white, and of varying degrees of competence. Sadly, by the time of the book's publication the Osprey was in drastic decline in Britain, its penchant for taking fish within sight of those charged with protecting fish ponds and lakes making it a target for widespread and relentless persecution.

Contemporary culture

Thankfully, we now live in far more enlightened times, and the Osprey's global distribution, combined with its recent recovery and iconic status in many countries, means that you do not need to look far to find references to the species. The United States National Football League team the Seattle Seahawks and Swansea-based Welsh Rugby Union team the Ospreys are just two of the numerous sporting teams named after them. The Osprey was officially chosen as the provincial bird of Nova Scotia, Canada, by an Act of the House of Assembly in 1994, and is also the provincial bird of Södermanland in Sweden. It has been featured on more than 50 postage stamps worldwide, including the Royal Mail's 2018 Special Stamp programme, which celebrated reintroduced species. The Bell Boeing V-22 Osprey, a United States tiltrotor military aircraft, is named after the species, and many companies use Osprey in their name. These include one that sells outdoor clothing and backpacks, another that produces luxury leather products, and even one that specialises in heavy transport lifting.

Below: The Bell Boeing V-22 Osprey is an American multi-mission, tiltrotor military aircraft.

Left: A roundabout on the outskirts of Oakham in Rutland features an impressive Osprey sculpture.

Below: Boat of Garten in Scotland famously describes itself as the 'Osprey village'.

In the UK, estate agents, restaurants, travel agents and even a leisure centre are all named after the Osprey. You can buy a pint of Rutland Osprey from the Grainstore Brewery in Rutland, while Golden Seahawk is brewed in Devon. In essence, the Osprey 'brand' is everywhere. This is particularly true in areas that have become synonymous with nesting Ospreys. Boat of Garten in the Scottish Highlands famously declared itself the Osprey village, but as the species continues to expand, other places might challenge this status. Oakham in Rutland, for example, has a beautiful Osprey carving on one of its roundabouts, in celebration of the successful Osprey translocation at nearby Rutland Water. This all points to the fact that the Osprey is an instantly recognisable bird; it is regarded as a purveyor of good news and perhaps a sign of how nature can recover when given a chance.

A flagship species

In the 1960s, when George Waterston made the visionary decision to publicise the location of the Osprey nest at Loch Garten (see page 72), he provided the opportunity for members of the British public to view breeding Ospreys for the first time. It also created a model for modern-day ecotourism that has been replicated across the UK and around the world. Fifty years later, the Osprey is an icon of conservation, an example of what can be achieved with a proactive approach. It has become a flagship species, a bird that can inspire the general public about nature and conservation issues like few others.

Over the years, 2.5 million people have made the pilgrimage to see the Ospreys at Loch Garten, but as the birds become more widespread across the UK, they are coming into contact with an even higher number of people. It is estimated that 125,000 people visit five Osprey-watching sites in Scotland every year, while more than 80,000 people visit the Bassenthwaite Ospreys in the Lake District each summer. At Rutland Water, 1,000 people go on special Osprey cruises aboard the *Rutland Belle* between late May and August, and in Dorset, the first four Poole Harbour Osprey cruises sold out within days of going on sale in summer 2017.

Below: Looking for Ospreys aboard the *Rutland Belle* at Rutland Water.

Osprey-related tourism is now vital in several parts of the country, contributing more than £2.2 million per year to the economy in Scotland, and more than £2 million in Cumbria. B&Bs, pubs and restaurants all directly benefit from the presence of breeding Ospreys, while Osprey cruises at Rutland Water and the nearby photography hide at River Gwash Trout Farm (see page 22) are excellent examples of how local businesses have embraced having Ospreys in the area. It is particularly significant that the photography hide is at a working trout farm. Prior to the construction of the hide, predation by Cormorants (*Phalacrocorax carbo*), Otters (*Lutra lutra*), Grey Herons (*Ardea cinerea*) and Ospreys was having a significant impact on fish stocks, but the income now generated by photographers more than offsets any such losses. In fact, the hide has become an integral part of the business, which owes much to the determination of owner Lawrence Ball and his colleague Jamie Weston in finding a positive solution to the problem. The hide means that everyone, including Ospreys, now benefits.

Above: A specially designed hide at River Gwash Trout Farm enables wildlife photographers to capture some stunning images of fishing Ospreys.

Below: Thousands of people visit Osprey-watching sites around the UK each year, greatly benefitting the rural economy in those places with Ospreys in the process.

Streamed to your home

Above: 4K cameras provide incredible views of the Dyfi Ospreys in Wales.

Today, modern technology means that you do not even need to leave the comfort of your home to enjoy watching Ospreys. Live images from nests in North America and across Europe are streamed online in high definition, giving viewers a remarkable insight into Osprey family life. It is particularly exciting to watch the footage around hatching time, when viewers are able to witness the moment a chick breaks free of its shell for the first time, and the tenderness with which it is fed by its mother a few hours later. The Ospreys featured on these cameras become celebrities in their own right, and some cameras have generated a considerable following. High-resolution 4K cameras installed by the Montgomeryshire Wildlife Trust at their Cors Dyfi nest provide some of the best footage of nesting Ospreys anywhere in the world, and the birds there have become very well known since being featured on the BBC's *Springwatch* programme. When you factor in that it is also possible to follow individual satellite-tracked Ospreys on their amazing migrations – often on a minute-by-minute basis – it is easy to understand how people become hooked on the Osprey story.

Reaching out

It is clear that Ospreys appeal to a much wider audience than many of our other, less charismatic species, and as such, they provide a means with which to inspire the general public about nature, as well as raise awareness of specific environmental issues. Visit any Osprey-watching site in the UK, and the awe and excitement that the birds generate are palpable. Encouragingly, this is also the case elsewhere on Osprey migration flyways. In recent years, Ospreys have helped link schoolchildren from three different continents – Europe, North America and Africa – during World Osprey Week. This international celebration of Osprey migration has linked students in areas where Ospreys spend the northern winter, such as Tanji in the Gambia, to places where they breed in Europe and North America. An associated environmental-education programme run at several schools in the Gambia has been a real hit with the students, some of whom have gone on to help out with beach clean-ups and surveys of wintering Ospreys as a result of their increased knowledge and interest in environmental issues. In fact, Tanji's football team is now called Osprey FC.

Outreach work carried out by the Rutland Osprey Project education team has been equally effective in the Rutland Water area, and students from schools in the two countries are now in regular contact, brought together by their shared interest in these wonderful birds. This just goes to show that wherever it occurs in the world, the Osprey is a bird that can ignite people's interest in nature and a desire to protect it.

Below: World Osprey Week is held in spring each year and encourages schools around the world to link up and to incorporate Ospreys into their studies.

Watching Ospreys

Today, the work of George Waterston and colleagues at Loch Garten in the 1960s (see page 71) has been replicated at many sites around the UK. With breeding Ospreys now present in many parts of the country, and dedicated Osprey visitor centres and purpose-built viewing hides, there has never been a better time to get out there and enjoy seeing this fantastic bird.

The first Ospreys are usually seen in the UK in early March. You can encounter a migrant Osprey just about anywhere, but estuaries, reservoirs and gravel pits are generally the best places to look.

Opposite: Ospreys generate huge interest when they return to the UK in the spring.

There are various public viewing sites around the country where you can see the birds once they are back at their nesting sites. It is essential to take your binoculars with you, but most places have telescopes to help you get an even better view of the nest, and knowledgeable and friendly staff to answer your questions. Many of the projects are supported by a team of volunteers – this is a great way to get involved in Osprey conservation.

Below: One of the watchpoints at Bassenthwaite Lake in Cumbria, which is visited by over 80,000 people hoping to see Ospreys each year.

Above: June and July, when chicks are growing fast and preparing to fledge, is a great time to visit Osprey watchpoints.

Chicks usually hatch in late May and early June, and so June and July are a great time to visit an Osprey viewing site. If you visit in early or mid-July, you may be lucky enough to witness the moment a chick takes to the air for the first time. Although migrant Ospreys rarely linger for long in the spring, autumn migration is often a more leisurely affair, and birds will sometimes hang around at favoured spots in August, September and even the early part of October. Juveniles may stay for several weeks if they find fish easy to catch. These birds can make for great viewing as they hone their hunting skills before continuing their journey.

It is always worth checking any Osprey you see for colour rings, which can be read in the field with a telescope or by zooming in on a good-quality photo of the bird. An image taken with a digital SLR camera with a 400mm lens should provide a good chance of reading the ring, providing the bird flies past relatively close. You can report any colour-ring sightings to the Roy Dennis Wildlife Foundation using a simple online form (roydennis.org/report-a-colour-ringed-osprey).

Scotland

Loch Garten, Inverness-shire

rspb.org.uk/lochgarten

More than 60 years after Ospreys first bred here, Loch Garten remains one of the prime Osprey-viewing locations in the UK. The RSPB visitor centre in the heart of Abernethy Forest provides excellent views of the long-established nest, and staff and volunteers are on hand to give you all the latest news.

Loch of the Lowes, Perthshire

scottishwildlifetrust.org.uk/reserve/loch-of-the-lowes

Like Loch Garten, the Scottish Wildlife Trust Reserve at Loch of the Lowes is synonymous with breeding Ospreys. A purpose-built viewing hide, just a short walk from the visitor centre, offers panoramic views across the loch to the Osprey nest on the opposite shore.

England

Bassenthwaite Lake, Cumbria

ospreywatch.co.uk

Ospreys returned to Cumbria in 2001, when a pair bred successfully at Bassenthwaite Lake near Keswick. They have continued to breed there each year since, and the nest can be seen from two viewpoints at Dodd Wood, run by the RSPB, the Forestry Commission and Lake District National Park. Some of the Lake District's most iconic peaks provide a spectacular backdrop from the upper viewpoint, a 20-minute walk along the forest road.

Foulshaw Moss, Cumbria

cumbriawildlifetrust.org.uk/reserves/foulshaw-moss

Ospreys first bred at Foulshaw Moss – a lowland raised peatbog of international importance south of Kendal – in 2014, and they now return each year. You can observe the birds from a special viewpoint on this Cumbria Wildlife Trust reserve.

Kielder Forest, Northumberland

kielderospreys.wordpress.com

Ospreys returned to breed in Kielder Forest in 2009, and the Northumberland Wildlife Trust coordinates a team of knowledgeable volunteers who share their expertise with visitors from a viewing point situated behind the Boat Inn at Kielder Waterside. Live images are relayed to the viewpoint as well as to Kielder Castle café. The Kielder Osprey Project is a partnership between the Northumberland Wildlife Trust, the Kielder Water and Forest Park Development Trust, and the Forestry Commission.

Poole Harbour, Dorset

birdsofpooleharbour.co.uk

rspb.org.uk/arne

The site of the second English Osprey translocation is a great place to view the birds. Although the release site is on private land with no public access, RSPB Arne is an excellent place to view the translocated birds after they have been released in August. There is also a chance of seeing passing or summering Ospreys anywhere in the harbour from March through to early October, and the Birds of Poole Harbour run Osprey cruises during August.

Rutland Water, Rutland

ospreys.org.uk

The Lyndon Visitor Centre forms the base of the Rutland Osprey Project, a partnership between the Leicestershire and Rutland Wildlife Trust and Anglian Water. Live images are relayed to the visitor centre, and two hides offer spectacular views of the Manton Bay nest. In addition, special Osprey cruises run on the *Rutland Belle* provide a way to see fishing Ospreys in action.

Wales

Cors Dyfi, Powys

dyfiospreyproject.com

Ospreys first nested at the Montgomeryshire Wildlife Trust's Cors Dyfi Reserve in 2011, and you can enjoy stunning views of the nest, along with live CCTV images, from the comfort of the purpose-built 360 Observatory, opened in 2014.

Glaslyn, Gwynedd

glaslynwildlife.co.uk

Ospreys have bred in the Glaslyn Valley each year since 2004. Glaslyn Wildlife operates a viewing point in a stunning location at Pont Croesor, with a backdrop of Snowdonia.

Photographing Ospreys

There are now several places where wildlife photographers have the opportunity to photograph fishing Ospreys at very close quarters from specially designed hides. The birds are often so close that any digital camera will give you a chance of capturing the magical moment an Osprey dives, but if you are a serious wildlife photographer, then a digital SLR with a 100–400mm is even better. Staff at the following sites will be able to provide other practical location-specific information. There are Osprey hides at at Rothiemurchus Fishery and also at nearby Aviemore Ospreys, both in Inverness-shire (rothiemurchus.net and aviemoreospreys.co.uk). There is also a photographic hide at a working trout farm at Horn Mill near Exton in Rutland, which provides the best opportunity to photograph fishing Ospreys in England (rivergwashtroutfarm.co.uk).

Above: Fishing in Rothiemurchus, Highland, Scotland.

Fu

Boo

Dennis
    Pub
Evans,
    Yea
Mackri
    Blo
Poole,
    His

On

There
Ospre
live st
tracki
view
websi
additi
check

Finnis
luom
Satell
websi

Rob
ospre
Satell

A

Sinc
at Bl
also
to p
exce

and
wor
hav
to K

all
has
me
and
to
dor

enc

Image credits

Bloomsbury Publishing would like to thank the following for providing photographs and for permission to reproduce copyright material. While every effort has been made to trace and acknowledge all copyright holders, we would like to apologise for any errors or omissions and invite readers to inform us so that corrections can be made in any future editions of the book.

Key t = top; l = left; r= right; tl = top left; tcl = top centre left; tc = top centre; tcr = top centre right; tr = top right; cl = centre left; c = centre; cr = centre right; b = bottom; bl = bottom left; bcl = bottom centre left; bc = bottom centre; bcr = bottom centre right; br = bottom right

AL = Alamy; FL= FLPA; G = Getty Images; NPL = Nature Picture Library; RS = RSPB Images; SS = Shutterstock, iStock = iS

Cover Page 1 kristianbell/G; **3** John Wright; **4** Geoff Harries; **5** SS; **6** tirc83/iS; **7** SCOTLAND: The Big Picture/NPL; **8** Andy Rouse/NPL; **9** John Wright; **10** l Jurgen & Christine Sohns/FL, r MikeLane45/iS; **11** l Iain Brownlee/AL, r Stephanie Jackson/AL; **12** SS; **13** SS; **14** John Wright; **15** Kevin Sawford/RS; **16** John Wright; **17** MikeLane45/iS; **18** cyricc/iS; **19** John Wright; **20** Mark Hamblin/RS; **21** Michael Cohen/Stringer/G; **22** Geoff Harries; **23** John Wright; **24** Andrew Mason/FL; **25** Ingo Arndt/NPL; **26** t John Wright, b SS; **27** John Wright; **28** John Wright; **29** John Wright; **30** John Wright; **31** Allen Creative/AL; **32** Dave Pressland/FL; **33** John Wright; **34** Â© Biosphoto, Juan-Carlos Munoz/Biosphoto/FL; **35** Â© Biosphoto, Juan-Carlos Munoz/Biosphoto/FL; **36** t Paul Hobson/NPL, b SCOTLAND: The Big Picture/NPL; **37** Larry Minden/FL; **38** LRWT/Rutland Osprey Project; **39** LRWT/Rutland Osprey Project; **40** LRWT/Rutland Osprey Project; **41** t LRWT/Rutland Osprey Project, b Tim Mackrill; **42** John Wright; **43** John Wright; **44** Michael Callan/FL; **45** John Wright; **46** John Wright; **47** SS; **48** Alan Gililand; **49** SS; **50** John Wright; **51** John Wright; **53** John Wright; **54** Imagebroker/FL; **55** Alan Gililand; **56** John Wright; **57** John Wright; **58** John Wright; **59** John Wright; **60** Jean-Claude Lehoucq; **61** John Wright; **62** John Wright; **63** Farid Lacroix; **64** John Wright; **65** John Wright; **66** John Wright; **67** John Wright; **68** Chris Gomersall/AL; **69** John Wright; **70** Antiqua Print Gallery/AL; **71** Charles Eric Palmer/photoscot.co.uk; **72** John Arnott/SOC archive; **73** LRWT/Rutland Osprey Project; **74** John Wright, b Tim Mackrill; **75** Andrew Harrington/NPL; **76** Andrew_Howe/iS; **77** Tony Hamblin/FL; **78** SS; **79** Tim Mackrill; **80** Tim Mackrill; **81** Tim Mackrill; **82** John Wright; **83** Forestry Commission England; **84** LRWT/Rutland Osprey Project; **85** Tim Mackrill; **86** LRWT/Rutland Osprey Project; **87** John Wright; **88** Will Watson/NPL; **89** David Tipling/2020VISION/NPL; **90** John Wright; **91** John Wright; **92** LRWT/Rutland Osprey Project; **93** John Wright; **94** Glaslyn Wildlife; **95** Montgomeryshire Wildlife Trust; **96** Raylipscombe/iS; **97** Aitor Galarza; **98** Adolfo Ventas; **99** SS; **100** Luis Palma; **101** Flavio Monti; **102** Denis Landenbergue; **103** Simon Kidner; **104** Geoff Harries; **105** SS; **106** Hulton Archive/Stringer/G; **107** duncan1890/G; **108** SS; **109** The Natural History Museum/AL; **110** drial7m1/G; **111** t Tim Mackrill, b LRWT/Rutland Osprey Project; **112** John Wright; **113** t John Wright, b Geoff Harries; **114** ; **115** Provo Primary School; **116** Andy Rouse/NPL; **117** Ray Kennedy/RS; **118** Mark Hamblin/G; **119** t Ray Wise/G, b Steve Gardner; **120** t Linda Lyon/G, b Cumbria Wildlife Trust; **121** t Roger Coulam/G, b SS; **122** t Paul Stammers, b Montgomeryshire Wildlife Trust; **123** t Gruff Owen, b SS.

Index